The Third Pillar of International Climate Change Policy

T0174150

During the negotiations in 2015 that led to the adoption of the Paris Agreement, one of the most contentious issues was the introduction of a dedicated provision in Article 8 on what is known as 'loss and damage'. The adoption of this new article, however, left many questions unanswered. What is the distinction between 'loss and damage', and 'adaptation'? What are the legal implications of the inclusion of loss and damage as an article in a legal treaty? How can financial assistance and compensation best be channelled to victims of climate change loss and damage? What gaps remain in the loss and damage governance system?

The Third Pillar of International Climate Change Policy: On 'Loss and Damage' after the Paris Agreement addresses these questions, and numerous others, and explores the present and future of loss and damage in the era of the Paris Agreement. This book provides an up-to-date analysis of 'loss and damage' which is often described as the third pillar of international climate change policy. It is based around four main themes: (i) insurance schemes, (ii) key gaps in loss and damage governance, including non-economic loss and damage and slow-onset events, (iii) legal aspects of loss and damage, and (iv) novel approaches to loss and damage.

The chapters in this book were originally published as a special issue of *Climate Policy*.

Morten Broberg is Professor of International Development Law and Honorary Jean Monnet Professor, the Faculty of Law, University of Copenhagen, Denmark.

Beatriz Martinez Romera (PhD, MSc, LLM) is Associate Professor of Law at the Center for International Law and Governance, the Faculty of Law, University of Copenhagen, Denmark. She has worked on environmental and climate change law and governance since 2010, in particular the regulation of greenhouse gas emissions from international aviation and maritime transport. She is the founder and manager of the Transatlantic Maritime Emissions Research Network (TRAMEREN) and member of the Bar Association of Madrid, Spain.

The Third Pillar of International Climate Change Policy

On 'Loss and Damage' after the Paris Agreement

Edited by
Morten Broberg and Beatriz Martinez Romera

LONDON AND NEW YORK

First published 2021
by Routledge
2 Park Square, Milton Park, Abingdon, Oxon, OX14 4RN

and by Routledge
605 Third Avenue, New York, NY 10158

Routledge is an imprint of the Taylor & Francis Group, an informa business

Introduction, Chapters 1, 3–5, 8 and 9 © 2021 Taylor & Francis
Chapter 2 © 2019 Morten Broberg. Originally published as Open Access.
Chapter 6 © 2019 Margaretha Wewerinke-Singh and Diana Hinge Salili. Originally published as Open Access.
Chapter 7 © 2019 Morten Broberg. Originally published as Open Access.

With the exception of Chapters 1, 6 and 7, no part of this book may be reprinted or reproduced or utilised in any form or by any electronic, mechanical, or other means, now known or hereafter invented, including photocopying and recording, or in any information storage or retrieval system, without permission in writing from the publishers. For details on the rights for Chapters 1, 6 and 7, please see the chapters' Open Access footnotes.

Trademark notice: Product or corporate names may be trademarks or registered trademarks, and are used only for identification and explanation without intent to infringe.

British Library Cataloguing-in-Publication Data
A catalogue record for this book is available from the British Library

ISBN13: 978-0-367-67668-1 (hbk)
ISBN13: 978-0-367-67669-8 (pbk)
ISBN13: 978-1-003-13227-1 (ebk)

Typeset in Myriad Pro
by codeMantra

Publisher's Note
The publisher accepts responsibility for any inconsistencies that may have arisen during the conversion of this book from journal articles to book chapters, namely the inclusion of journal terminology.

Disclaimer
Every effort has been made to contact copyright holders for their permission to reprint material in this book. The publishers would be grateful to hear from any copyright holder who is not here acknowledged and will undertake to rectify any errors or omissions in future editions of this book.

Contents

Citation Information

The chapters in this book, except Chapter 7, were originally published in the *Climate Policy*, volume 20, issue 6 (October 2020). Chapter 7 was published in volume 20, issue 5 of the same journal. When citing this material, please use the original page numbering for each article, as follows:

Chapter 1
Insurance schemes for loss and damage: fools' gold?
Linnéa Nordlander, Melanie Pill and Beatriz Martinez Romera
Climate Policy, volume 20, issue 6 (October 2020) pp. 704–714

Chapter 2
Parametric loss and damage insurance schemes as a means to enhance climate change resilience in developing countries
Morten Broberg
Climate Policy, volume 20, issue 6 (October 2020) pp. 693–703

Chapter 3
Non-economic loss and damage: lessons from displacement in the Caribbean
Adelle Thomas and Lisa Benjamin
Climate Policy, volume 20, issue 6 (October 2020) pp. 715–728

Chapter 4
Loss and damage in the IPCC Fifth Assessment Report (Working Group II): a text-mining analysis
Kees van der Geest and Koko Warner
Climate Policy, volume 20, issue 6 (October 2020) pp. 729–742

Chapter 5
Loss & damage from climate change: from concept to remedy?
Meinhard Doelle and Sara Seck
Climate Policy, volume 20, issue 6 (October 2020) pp. 669–680

Chapter 6
Between negotiations and litigation: Vanuatu's perspective on loss and damage from climate change
Margaretha Wewerinke-Singh and Diana Hinge Salili
Climate Policy, volume 20, issue 6 (October 2020) pp. 681–692

Chapter 7

Chapter 8

Chapter 9

For any permission-related enquiries please visit:
http://www.tandfonline.com/page/help/permissions

Contributors

Lisa Benjamin Schulich School of Law, Dalhousie University, Halifax, Canada.

Adrian Martínez Blanco Institute for Advanced Sustainability Studies (IASS) e.V., Potsdam, Germany. La Ruta del Clima, San José, Costa Rica.

Morten Broberg Faculty of Law, University of Copenhagen, Denmark.

Meinhard Doelle Schulich School of Law, Dalhousie University, Halifax, Canada.

Kees van der Geest United Nations University Institute for Environment and Human Security, Bonn, Germany.

Linnéa Nordlander Faculty of Law, University of Copenhagen, Denmark.

Mark Pelling Overseas Development Institute, London, UK.

Melanie Pill Fenner School of Environment and Society, Australian National University, Canberra, Australia.

Erin Roberts Kings College London, UK. Overseas Development Institute, London, UK.

Beatriz Martinez Romera Faculty of Law, University of Copenhagen, Denmark.

Diana Hinge Salili Pacific Centre for Environment and Sustainable Development, University of the South Pacific, Suva, Fiji.

Sara Seck Schulich School of Law, Dalhousie University, Halifax, Canada.

Adelle Thomas Climate Analytics, Berlin, Germany.

Patrick Toussaint Law School, University of Eastern Finland, Joensuu, Finland. Institute for Advanced Sustainability Studies (IASS) e.V., Potsdam, Germany.

Koko Warner United Nations University Institute for Environment and Human Security, Bonn, Germany.

Margaretha Wewerinke-Singh Grotius Centre for International Legal Studies, Leiden, the Netherlands. Pacific Centre for Environment and Sustainable Development, University of the South Pacific, Suva, Fiji.

Loss and Damage after Paris: All Talk and No Action?

Morten Broberg and Beatriz Martinez Romera

Introduction

During the negotiations that led to the adoption of the Paris Agreement, one of the most contentious issues was the introduction of a dedicated provision in Article 8 on what is known as 'loss and damage'. The adoption of this new article, however, left many questions unanswered. What is the distinction between loss and damage, and adaptation? What are the legal implications of the inclusion of loss and damage as an article in a legal treaty? How can financial assistance and compensation best be channelled to victims of climate change loss and damage? What gaps remain in the loss and damage governance system?

Such questions, and numerous others, prompted a group of Nordic researchers to organise a workshop outside Copenhagen in May 2018 under the title 'Climate Change Adaptation and Loss & Damage after Paris – Bridging Different Levels of Governance'. The workshop focussed on bridging different levels of governance to address the impacts of climate change, and also on examining how the new provision on loss and damage affects the legal construction of the UN climate change treaty regime provisions on both mitigation and adaptation (Broberg, 2020a). This book, which includes contributions beyond the workshop participants, builds on these discussions, exploring the present and future of loss and damage in the era of the Paris Agreement (Broberg and Martinez Romera, 2020).

In what follows, we first present a short history behind the inclusion of loss and damage in the UN climate change treaty regime, before turning to definitional issues. We thereupon summarise key insights from the nine chapters of the book, based around four main themes: (i) insurance schemes, (ii) key gaps in loss and damage governance, including non-economic loss and damage and slow-onset events, (iii) loss and damage litigation, and (iv) novel approaches to loss and damage. In the final section, we present the picture that the nine chapters paint when viewed together – and end on a note of hope.

Loss & Damage in the UN Climate Change Treaty Regime

The inclusion of Article 8 in the Paris Agreement (UNFCCC, 2015) meant the conclusion of a longstanding process for the recognition of loss and damage in the climate change regime. This process started in 1991 when the Alliance of Small Island States (AOSIS) ignited discussions with a proposal for the introduction of a mechanism to address climate change loss and damage (Intergovernmental Negotiating Committee for a Framework Convention on Climate Change (INC), 1991) during the negotiations that led to the adoption of the United Nations Framework Convention on Climate Change (UNFCCC) in 1992. Although the proposal was not included in the final text, the idea of a mechanism to deal with loss and damage had been sown. Different ways of addressing climate change-induced loss and damage were subsequently examined, including in the 2007 Bali Action Plan (UNFCCC, 2008). However, it was not until the 16th Conference of the Parties (COP 16) in Cancun in 2010 that a work programme was established to consider approaches 'to address loss and damage associated with climate change impacts in developing countries that are particularly vulnerable to the adverse effects of climate change' (UNFCCC, 2011). At COP 19, a major breakthrough came with the establishment of the Warsaw International Mechanism for Loss and Damage (WIM), which aims 'to address loss and damage associated with impacts of climate change, including extreme events and slow onset events, in developing countries that are particularly vulnerable to the adverse effects of climate change', and the establishment of an Executive

Committee (ExCom) to guide the implementation of functions of the WIM through an initial two-year work plan (UNFCCC, 2014). At the time of writing, the most recent substantive step came in 2015, as the WIM was followed up at COP 21 in Paris with a dedicated provision on loss and damage included in Article 8 of the Paris Agreement. This ensured that loss and damage was given a formal platform within the UN climate change treaty regime, but a number of developed countries were only willing to accept this on the condition that it would 'not involve or provide a basis for any liability or compensation' as was therefore expressly set out in paragraph 51 of decision 1/CP.21.

Article 8(4) provides a non-exhaustive list of eight 'areas of cooperation and facilitation to enhance understanding, action and support' through which the parties to the Paris Agreement may seek to avert, minimise and address loss and damage. While some of these areas of cooperation and facilitation may easily be categorised as loss and damage, such as Article 8(4)(g) concerning non-economic losses, others, such as Article 8(4)(a) concerning early warning systems, (b) concerning emergency preparedness, (e) concerning comprehensive risk assessment and management, and (h) concerning resilience of communities, livelihoods and ecosystems, could just as well have been labelled 'adaptation'. Indeed, as observed by Broberg (2020b) in the chapter 'Parametric loss and damage insurance schemes as a means to enhance climate change resilience in developing countries', at the very first COP in 1995, insurance was categorised as an 'adaptation measure' whereas today insurance is also considered a measure under loss and damage. This type of 'displacement' of provisions is, in itself, bound to give rise to uncertainty with regard to the delimitation between 'loss and damage' and 'adaptation'.

Working Towards Defining Loss and Damage

Neither the Paris Agreement nor the UN climate change treaty regime provides a formal definition of 'loss and damage', and both practitioners and academics apply diverging definitions which can broadly be divided into three different groups (see Figure 1).

The three most common ways of distinguishing 'adaptation' and 'loss and damage'

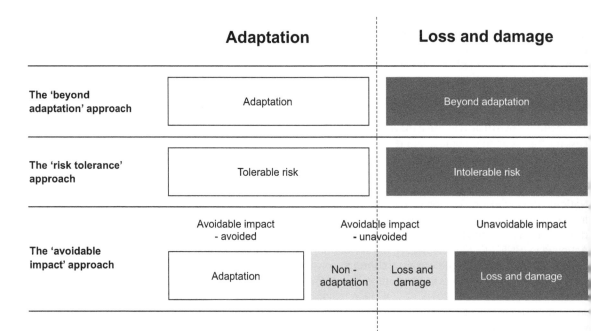

According to the first such definition, loss and damage covers measures that address the impacts of climate change that are 'residual' to mitigation and adaptation. In other words, if we pool 'insufficient mitigation' with 'inadequate adaptation' we will be left with loss and damage. Constructing loss and damage in this way is referred to as the 'beyond adaptation approach' (Mechler et al., 2019).

The second definition focuses on what may be considered to be the 'tolerable risk'. Here adaptation is about keeping risks within the range of what is perceived as 'tolerable', whereas loss and damage is a response to risks that cannot be kept within that range (Dow et al., 2013).

A third definition defines loss and damage by distinguishing between climate change impacts that are 'avoidable', 'unavoidable', or 'unavoided'. Thus, if it is impossible to adapt to an impact so that it becomes unavoidable, it will fall in the loss and damage category. For impacts that are avoidable, it is necessary to distinguish between those that are avoided and those that are not. If it is possible to adapt to an avoidable impact so that it is avoided, this is a case of adaptation. However, if an avoidable impact is not avoided, it is unclear from this definition whether it is to be categorised as (non) adaptation, or as loss and damage (van der Geest and Warner, 2015).

To illustrate the differences between the three above definitions, consider the case of a small mountain state that relies on drinking water from high-altitude glaciers. Unfortunately, the rising temperatures cause the glaciers to retreat – giving rise to serious disruptions to the water supply. The small mountain state therefore decides to address this challenge by laying water tubes to source those towns and villages which will otherwise be without water supplies; this measure will constitute a form of adaptation. Now assume that the state is incapable of building a network of tubes that will enable it to supply all its citizens – meaning that some towns and villages will necessarily have to be abandoned. If this incapacity is due to technical reasons, it falls into the loss and damage category irrespective of which of the three definitions we apply.

If, however, it were technically feasible to source all towns and villages with water, but the costs of this would exceed the economic capacity of the small mountain state, the decision of which towns and villages shall be sourced and which shall be abandoned will, essentially, be a political one regarding how to spend the scarce resources. Applying the third definition, where it is a political decision not to (fully) adapt, it is not clear whether the specific abandonment of certain towns and villages ('unavoided loss') should be categorised as '(non) adaptation' or as 'loss and damage'. The example thus illustrates the difficulties of drawing a clear distinction between adaptation and loss and damage.

Returning to the second definition which focuses upon the 'tolerable risk' (ie. where loss and damage is a response to risks that cannot be kept within the range of the tolerable), laying water tubes to selected towns and villages is a way of ensuring that risks remain tolerable (ie. adaptation). By contrast, abandoning some towns and villages to water famine is unlikely to be considered a tolerable risk and therefore falls into the loss and damage category. However, determining what is 'tolerable risk' is rather subjective, as it is based on the value judgments of those faced with the actual climate change impacts – meaning that it does not provide a clear and objective distinction between adaptation and loss and damage.

The different definitions of loss and damage may not only lead to different categorisations of cases as either adaptation or loss and damage. They may also lead to different priorities within the field of loss and damage (Broberg, 2020c).

As will be clear from the above, none of the three definitions is fully workable or clear-cut. Doelle and Seck (2020) have pointed to a different distinctions between adaptation, on the one hand, and loss and damage, on the other, where the focus is upon whether harm has been caused by human-induced climate change itself. Accepting this way of defining loss and damage means that adaptation will be about avoiding the occurrence of harm, whereas loss and damage will be about addressing the harm that occurs, whether or not it could have been reduced by adaptation. This approach also seems to be in line with Article 8(3) of the Paris Agreement, which merely refers to 'loss and damage associated with the adverse effects of climate change'.

This book

This book, building on the discussions at the 2018 Nordic researchers' workshop in Copenhagen and also including contributions beyond the workshop participants, explores the present and future of loss and damage in the era of the Paris Agreement. Not only is the Paris Agreement's provision on loss and damage novel, it is also highly disputed, as is clearly reflected in the chapters that make up the present book.

When a new substantive provision is introduced into a pre-existing legal regime, two crucial questions arise: What can it be used for, and what gaps does it create or leave unaddressed? These are the two key questions that the chapters of this book address with respect to Article 8 of the Paris Agreement. In investigating the answers to these two questions, the nine contributions centre around four different themes regarding the post-Paris Agreement governance of loss and damage.

Insurance Schemes

Much of the discussion regarding loss and damage has focussed upon insurance as an important means of addressing loss and damage caused by climate change, and so insurance also plays a prominent role in this book.

Nordlander, Pill and Martinez Romera (2020) critically review the limitations of what they term 'active insurance schemes for loss and damage'. They do so with regard to the principles of common but differentiated responsibilities and respective capabilities, and intergenerational equity, enshrined in the climate change regime, along with economic and gender inequality, and human mobility. The authors conclude that, despite their popularity among policy makers, the potential for insurance schemes to deliver appropriate financial responses to loss and damage is limited. Ultimately, they find that insurance schemes are fundamentally misaligned with founding principles of the international climate change regime. Further, insurance will become less relevant over time and cannot address the full range of anticipated and incurred losses and damages, so policy makers are encouraged to consider how to overcome these challenges through innovative approaches to insurance design and additional sources of finance. In this regard, a number of mechanisms to mobilise loss and damage finance have been proposed, such as the climate damages tax on fossil fuel companies discussed by Wewerinke-Singh and Hinge Salili (2020).

One possible mechanism is so-called parametric insurance, which differs from conventional insurance schemes since pay-outs are not based on an assessment of the actual post-event losses, but are instead triggered by certain pre-defined parameters being met. Using parametric insurance means that when an insurance event occurs, in principle it is relatively straightforward to establish objectively whether the conditions for payment have been fulfilled and, if they have, it is equally easy to calculate the size of payments due without having to first establish the actual losses. This means that payment can be made swiftly which may be particularly attractive in a developing country context. Broberg (2020b) examines three existing multi-country parametric risk pooling schemes in order to consider the pros and cons of using such schemes in a loss and damage context. He shows that parametric insurance schemes may provide a useful tool to this end, provided that considerable care is put into designing them – and even then it is important to be acutely aware of their limitations.

Key Gaps in Loss and Damage Governance: Non-economic Loss and Damage and Slow-onset Events

In addressing loss and damage, efforts are often primarily circumscribed to insurance approaches that target economic losses: non-economic loss and damage (including knowledge, social cohesion, identity, or cultural heritage) as well as slow onset events are largely left unaddressed. For small island developing states (SIDS) in the Caribbean, the 2017 hurricane season brought the issue of loss and damage to the fore when two category five storms passed through the region within two weeks of each other, resulting in severe damage on many islands. Media and political attention focussed on damages to the infrastructure and losses suffered by different industries. In their article, Thomas and Benjamin (2020) show that this focus on economic costs obscures the significant non-economic loss and damage – such as effects on health, sense of place and community cohesion – that may be inflicted by natural disasters linked to climate change. They argue that there is a need for robust national policies to address these impacts. Based on a study of how the 2017-hurricanes led to prolonged displacement of the entire population on The Bahamas' Ragged Island, the authors highlight the need for SIDS to develop national policies on non-economic loss and damage and to ensure that these countries have the national capacities to identify, assess, and report on loss and damage – in all of its forms – to facilitate access to support.

In this connection, the role of research in addressing slow-onset events and non-economic losses is emphasised in the article by van der Geest and Warner (2020). These authors observe that the contribution from Working Group II to the Fifth Assessment Report of the Intergovernmental Panel on Climate Change (IPCC) primarily associates loss and damage with extreme weather events and economic impacts, and that the Report treats it primarily as a future risk, whereas present-day loss and damage from slow-onset processes and non-economic losses receive much less attention. The analysis also shows that the IPCC's Fifth Assessment Report has more to say about losses and damages in high-income regions than in regions that are most at risk, such as small island states and least developed countries. The authors point out that people-centred research by social scientists is crucial for enhancing understanding of what loss and damage means in the real world and that, already today, loss and damage is a reality for vulnerable people in climate hotspots. The lack of empirical research on loss and damage from slow-onset processes and non-economic loss and damage in vulnerable countries needs to be rectified. Amongst others, the authors recommend that funding agencies should concentrate research support in these knowledge areas. Also, there is a role to play by Working Group II to the IPCC's Sixth Assessment Report (under preparation at the time of writing), which should be more alert to literature on loss and damage from slow-onset processes and non-economic loss and damage, particularly in vulnerable countries in the Global South.

Legal Aspects of Loss and Damage

In this book, several of the chapters coalesce around the legal aspects of loss and damage; including loss and damage litigation, in the context of whether Article 8 of the Paris Agreement can be used to force action in order to address loss and damage from climate change.

In their contribution, Doelle and Seck (2020) conceptualise loss and damage through the concept of harm and argue that litigation may work as a remedy and a tool to challenge current domestic legal systems to provide effective responses to loss and damage. The authors consider a number of entities (states, sub-national government actors and non-state actors, from individuals to organisations and communities) who might bring claims for loss and damage. The claimants might seek remedies according to the harm incurred and beyond monetary compensation, following the general trend of climate litigation. The matter against whom the claims may be addressed (states, state actors, state owned enterprises, international organisations, and private actors) is closely linked to the actionable wrongs at play, as well as the nature of the harm, the plaintiff, the legal system involved, and the remedies sought. The actionable wrongs can take a number of forms from government inaction to private actors misleading governments. The authors lucidly show that both domestic and international legal systems will be faced with claims for loss and damage, with different legal systems taking their own unique approach. This, in turn, will create a patchwork of venues with a different mix of eligible claimants, respondents, remedies, and actionable wrongs. Whether this patchwork will develop into a cohesive whole that offers appropriate remedies to all legitimate claimants remains to be seen.

Wewerinke-Singh and Hinge Salili (2020) examine the question of litigation using the Pacific island state of Vanuatu as an example. They argue that some developed states are deliberately blocking the operationalisation of the Paris Agreement regime in relation to loss and damage finance thereby increasing the potential importance of legal action to address loss and damage. The authors discuss the possibilities for legal action to seek redress for climate loss and damage through action against states under international law, and against fossil fuel companies under domestic law. The added value of climate litigation to promote public debate over difficult issues is highlighted. The authors show that the issue of compensation for climate loss and damage is best addressed at the multilateral level, and they consider how the two processes of litigation and negotiation may interact with each other and inspire more far-reaching action. With regard to financing, they point to the introduction of a climate damage tax on fossil fuel companies as a particularly promising option and they observe that using legal action may contribute to a comprehensive multilateral agreement on loss and damage with enforceable provisions on climate finance. In this connection, legal action represents a means to improve the position of climate-vulnerable states in multilateral negotiations on loss and damage finance.

In 'Interpreting the UNFCCC's provisions on 'mitigation' and 'adaptation' in light of the Paris Agreement's provision on 'loss and damage', Broberg (2020a) examines how the introduction of Article 8 on loss and damage in the Paris Agreement affects the construction of provisions on 'mitigation' and 'adaptation' as established within the UNFCCC treaty framework. He shows that the establishment of loss and damage at treaty level has created a legal basis for finding 'responsibility' for adverse consequences that can be attributed to the failure to fulfil UNFCCC obligations as laid down in the provisions on mitigation and adaptation. This, he argues, strengthens the legal basis for pursuing remedies aimed at reparation for these consequences, such as the establishment of climate change funds and of insurance solutions. Moreover, he demonstrates that prior to establishing loss and damage at treaty level, loss and damage issues and measures (such as the Warsaw International Mechanism) were treated in legal terms within the framework of adaptation. However, with the adoption of the Paris Agreement, loss and damage has been given its own legal basis and therefore loss and damage issues and measures must henceforth be treated within this new framework.

Novel Approaches to Loss and Damage

The potential of human rights law, and human rights approaches to provide remedies and thereby fill gaps in the field of climate change law and litigation where other areas of the law do not, has been broadly acknowledged in the literature (Savaresi and Auz, 2019). This potential with specific regard to loss and damage is explored in the contribution by Toussaint and Martínez Blanco (2020). Climate change has been labelled the human rights challenge of the twenty-first century and, in particular, loss and damage resulting from climate change poses a severe threat to the human rights of affected communities. The authors review the extent to which the international response to climate change under the UNFCCC has taken human rights into account, and find an insufficient level of integration of human rights in international climate governance. Framing loss and damage through the lenses of human rights and obligations, and adopting a human-rights based approach (HRBA) could strengthen the international response to loss and damage. A HRBA to loss and damage could remedy the framing of loss and damage in abstract, state-centric terms as a developing country issue, and instead put the spotlight on the fundamental human rights of the individual, including consideration of the intersectionality of loss and damage impacts, with, among others, questions of race, gender, class, age, and economic well-being. It could also empower victims as 'active participants' in decisions that concern their lives and livelihoods and promote the institutionalisation of cooperation between states and those most affected (Broberg and Sano, 2018). Among the ways in which this HRBA to loss and damage can be operationalised, there is an opportunity for the WIM to develop human rights guidelines for loss and damage policies and actions, as well as guidelines for conducting human rights impact assessments, and to set up a specialised body to monitor compliance.

Potential is also found in conceptualising loss and damage as a development crisis. Roberts and Pelling (2020) emphasise the importance of taking a broader societal approach to climate change. Thus, until now, climate change has predominantly been considered to be an environmental problem, often neglecting the social, political, cultural, and ethical dimensions of the issue. By conceptualising climate change as a development crisis, opportunities for transformation to address the root causes of loss and damage will emerge. The authors point to the emergence of a broader policy framework of loss and damage as a platform for transformation, and examine the need for a systematic assessment of the ways in which transformation might be deployed to best meet the aims of this broader loss and damage policy framework to avert, minimise and address the impacts of climate change that are not avoided. The two authors demonstrate the range of existing interpretations of transformation that can, or have been, applied to shape the broader policy framework of loss and damage policy and practice. They identify three approaches:

- Transformation as intensification (meaning types of transformation that reinforce rather than challenge the status quo).
- Transformation as extension (referring to the situation where the limits of established adaptive capacity are reached).
- Transformation as liberation (adopting development pathways that address the root causes of vulnerability).

Based on an analysis of these three approaches, the authors find that transformation as liberation offers the widest range of policy opportunities for the broader policy framework of loss and damage to meet the goals of equitable and sustainable development; and they go on to offer a set of potential, enabling factors and recommendations to help transition transformation policy into practice, as well as highlighting the role of global processes in facilitating transformation.

Conclusion

With the adoption of the 2015 Paris Agreement, the notion of loss and damage was given a formal platform within the UN climate change treaty regime. However, whereas Article 8 of the Agreement provided the bones for a loss and damage scheme there was still an obvious need to put flesh to these bones. At the time of writing – 2020 – this continues to be the case as is clearly reflected in the nine chapters of this book.

Thus, whereas all nine chapters acknowledge that the introduction of loss and damage in Article 8 of the Paris Agreement has been of material significance, the chapters also paint a clear picture that, until now, Article 8 has only been able to provide an incomplete loss and damage scheme. Consequently, it would not be unreasonable to assert that, so far, Article 8 has been 'all talk and no action'.

However, the chapters may also infuse some fresh hope, since several point to innovative ways of overcoming the significant gaps in the contemporary loss and damage scheme. Even though the solutions proposed are currently dwarfed by the size of the challenges, they do point to possible next steps toward a more effective regime to address loss and damage from the impacts of climate change.

Acknowledgments

This book has been based on 'Loss and Damage after the Paris Agreement' – a special issue of *Climate Policy* (volume 20, issue 6). The two co-editors would like to thank the fourteen authors for their contributions. They would also like to express their heartfelt gratitude to Dr. Joanna Depledge, editor of *Climate Policy*, for her work and indispensable role in the creation of the special issue; this work is also clearly reflected in the present book. Further, the co-editors are thankful to the Joint Committee for Nordic Research Councils in the Humanities and Social Sciences (NOS-HS) for funding the workshop 'Climate Change Adaptation and Loss & Damage after Paris – Bridging Different Levels of Governance', which was the origin of the ideas that are now published in this book.

Bibliography

Broberg, M. (2020a). Interpreting the UNFCCC's provisions on 'mitigation' and 'adaptation' in light of the Paris Agreement's provision on 'loss and damage'. *Climate Policy*, 20(5), 527–533, doi:10.1080/14693062.2020.1745744

Broberg, M. (2020b). Parametric loss and damage insurance schemes as a means to enhance climate change resilience in developing countries. *Climate Policy*, 20(6), 693–703. doi:10.1080/14693062.2019.1641461

Broberg, M. (2020c). The third pillar of international climate change law: Explaining 'loss and damage' after the Paris Agreement. *Climate Law*, 10(2), 211–223. doi:10.1163/18786561-01002004

Broberg, M., & Martinez Romera, B. (2020). Loss and damage after Paris: more bark than bite? *Climate Policy*, 20(6), 661–668. doi:10.1080/14693062.2020.1778885

Broberg, M., & Sano, H.-O. (2018). Strengths and weaknesses in a human rights-based approach to international development – an analysis of a rights-based approach to development assistance based on practical experiences. *The International Journal of Human Rights*, 22(5), 664–680. doi:10.1080/13642987.2017.1408591

Doelle, M., & Seck, S. (2020). Loss & damage from climate change: from concept to remedy? *Climate Policy*, 20(6), 669–680. doi:10.1080/14693062.2019.1630353

Dow, K. et al. (2013). Limits to adaptation. *Nature Climate Change*, 305, 305–306.

Intergovernmental Negotiation Committee for a Framework Convention on Climate Change Working Group II. (1991). *Vanuatu: Draft annex relating to Article 23 (Insurance) for inclusion in the revised single text on elements relating to mechanisms (A/AC.237/WG.II/Misc.13) submitted by the Co Chairmen of Working Group II*. Retrieved from: https://unfccc.int/sites/default/files/resource/docs/a/wg2crp08.pdf

Mechler, R. et al. (2019). Science for loss and damage: Findings and propositions, in *Loss and Damage from Climate Change: Concepts, Methods and Policy Options* (R. Mechler et al., eds.), Springer, 3.

Nordlander, L., Pill, M., & Martinez Romera, B. (2020). Insurance schemes for loss and damage: Fools' gold? *Climate Policy*, 20(6), 704–714. doi:10.1080/14693062.2019.1671163

Roberts, E., & Pelling, M. (2020). Loss and damage: An opportunity for transformation? *Climate Policy*, 20(6), 758–771. doi:10.1080/14693062.2019.1680336

Savaresi, A., & Auz, J. (2019). Climate change litigation and human rights: Pushing the boundaries. *Climate Law*, 9(3), 244–262. doi:10.1163/18786561-00903006

Thomas, A., & Benjamin, L. (2020). Non-economic loss and damage: Lessons from displacement in the Caribbean. *Climate Policy*, 20(6), 715–728. doi:10.1080/14693062.2019.1640105

Toussaint, P., & Martínez Blanco, A. (2020). A human rights-based approach to loss and damage under the climate change regime. *Climate Policy*, 20(6), 743–757. doi:10.1080/14693062.2019.1630354

UNFCCC. (2008). *FCCC/CP/2007/6/Add.1 Report of the Conference of the Parties on its thirteenth session, held in Bali from 3 to 15 December 2007 Addendum Part Two: Action taken by the Conference of the Parties at its thirteenth session.* Retrieved from: https://unfccc.int/resource/docs/2007/cop13/eng/06a01.pdf

UNFCCC. (2011). *FCCC/CP/2010/7/Add.1 Conference of the Parties Report of the Conference of the Parties on its sixteenth session, held in Cancun from 29 November to 10 December 2010 Addendum Part Two: Action taken by the Conference of the Parties at its sixteenth session.* Retrieved from: https://unfccc.int/resource/docs/2010/cop16/eng/07a01.pdf

UNFCCC. (2014). *FCCC/CP/2013/10/Add.1 Conference of the Parties Report of the Conference of the Parties on its nineteenth session, held in Warsaw from 11 to 23 November 2013 Addendum Part two: Action taken by the Conference of the Parties at its nineteenth session.* Retrieved from: https://unfccc.int/resource/docs/2013/cop19/eng/10a01.pdf

UNFCCC. (2015). *The Paris Agreement.* Retrieved from: https://unfccc.int/sites/default/files/english_paris_agreement.pdf

UNFCCC. (2016). *FCCC/CP/2015/10/Add.1 Conference of the Parties on its twenty-first session, held in Paris from 30 November to 13 December 2015 Addendum Part two: Action taken by the Conference of the Parties at its twenty-first session.* Retrieved from: https://unfccc.int/resource/docs/2015/cop21/eng/10a01.pdf

UNFCCC. (2018). *FCCC/CP/2018/10 Report of the Conference of the Parties on its twenty-fourth session, held in Katowice from 2 to 15 December 2018.* Retrieved from: https://unfccc.int/process-and-meetings/conferences/katowice-climate-change-conference-december-2018/sessions-of-negotiating-bodies/cop-24

UNFCCC. (2019). *FCCC/CP/2019/13 Report of the Conference of the Parties on its twenty-fifth session, held in Madrid from 2 to 15 December 2019.* Retrieved from: https://unfccc.int/documents?f%5B0%5D=conference%3A4252

UNFCCC. (2019). *FCCC/PA/CMA/2019/L.7 Warsaw International Mechanism for Loss and Damage associated with Climate Change Impacts.* Retrieved from: https://unfccc.int/sites/default/files/resource/cma2019_L07_adv_WIM.pdf

UN General Assembly. (1992). United Nations Framework Convention on Climate Change (UNFCCC). (Adopted 9 May 1992. Entry into force 21 March 1994). 1771 UNTS 107. Retrieved from: https://unfccc.int/resource/docs/convkp/conveng.pdf

van der Geest, K., & Warner, K. (2015). Vulnerability, coping and loss and damage from climate events. In A. Collins et al., (Eds.), *Hazards, Risks and Disasters in Society* (pp. 121–144). Elsevier.

van der Geest, K., & Warner, K. (2020). Loss and damage in the IPCC Fifth Assessment Report (Working Group II): A text-mining analysis. *Climate Policy*, 20(6) 729–742. doi:10.1080/14693062.2019.1704678

Wewerinke-Singh, M., & Salili, D. H. (2020). Between negotiations and litigation: Vanuatu's perspective on loss and damage from climate change. *Climate Policy*, 20(6), 681–692. doi:10.1080/14693062.2019.1623166

Insurance schemes for loss and damage: fools' gold?

Linnéa Nordlander ⓘ, Melanie Pill ⓘ and Beatriz Martinez Romera ⓘ

ABSTRACT

Insurance schemes are a widely supported form of finance mechanism to address climate change-induced loss and damage, and are part of the Warsaw International Mechanism for Loss and Damage. This paper reviews active insurance schemes for loss and damage by exploring existing critiques. Novel insights into the fundamental challenges that insurance schemes face are then examined, in particular in the context of common but differentiated responsibilities and respective capabilities, intergenerational equity, economic and gender inequality, and human mobility. The analysis concludes that, despite their popularity among policy makers, insurance schemes seem ill-suited to address the full range of loss and damage. Therefore, pursuing these schemes, without backstopping from international finance, might undermine the objective of responding to loss and damage in a comprehensive manner. Consequently, it may be advisable for policy makers to consider how to overcome the apparent challenges in order to 'insure the uninsurable'.

Key policy insights

- Existing insurance schemes are ill-suited to fully respond to climate change loss and damage, given the increased frequency and severity of sudden onset events, slow onset events, and non-economic losses and damages.
- Insurance schemes fail to align with principles enshrined in the climate change regime, in particular the principle of common but differentiated responsibilities and respective capabilities (CBDR-RC) and intergenerational equity.
- Insurance products do not take economic inequality or gender considerations into account and loss and damage stemming from human mobility does not lend itself to insurance solutions as currently conceived, in certain circumstances.
- If insurance continues to be pursued as a response to loss and damage, it requires a major overhaul with innovative approaches.
- Policy makers must consider sourcing new and additional finance, reflecting the principle of CBDR-RC.

1. Introduction

As the number and severity of climate change impacts rises, attention to loss and damage in international climate change fora has increased, and with it, the pressing need to find financial mechanisms to deal with climate change harm. The potential of insurance schemes to respond to this need had been explored at the regional level through insurance pools in developing countries long before their formal inclusion as part of the Warsaw International Mechanism for Loss and Damage (WIM). Insurance schemes have garnered support as financial tools for mitigating loss and damage (Lees, 2017),[1] but certain issues remain unaddressed, namely the lack of continued and secure finance, the inappropriateness of insurance to meet context-specific needs, and responsibility avoidance.

ⓑ Supplemental data for this article can be accessed https://doi.org/10.1080/14693062.2019.1671163

This paper focuses on the potential of insurance schemes to address climate-related loss and damage, with the aim of contributing to the existing literature on finance instruments in a three-fold way. First, the paper consolidates existing work – primarily stemming from policy papers, reports, academic literature, and legal sources – identifying design options for insurance schemes as a financial means of addressing climate change loss and damage (Blampied, 2016; Mahul, Boudreau, Lane, Beckwith, & White, 2011; Schäfer, Waters, Kreft, & Zissener, 2016; Whalley, 2016). Second, the paper provides an overview of the existing insurance schemes used to respond to climate change impacts and draws on recent case studies to evaluate their effectiveness. The final section analyses the unsuitability of insurance schemes in the context of sudden and slow onset events and non-economic loss and damage (NELD), and adds to the existing literature by critically analyzing insurance schemes in light of: (i) the principle of common but differentiated responsibilities and respective capabilities (CBDR-RC); (ii) intergenerational equity; (iii) economic inequality; (iv) gender considerations; and (v) human mobility.

2. Insurance for loss and damage and summary case studies

Calls for the establishment of a loss and damage mechanism to respond to the adverse impacts of climate change were made as early as 1991 (Roberts & Huq, 2015), and have since gained growing attention until its inclusion in the Paris Agreement (Article 8; UNFCCC, 2015). However, a crucial question remains over how to support a monetary response to loss and damage in an already underfunded climate change finance system. While a number of options have been discussed, much of the debate surrounds the use of insurance schemes.

The 1991 proposal of the Alliance of Small Island States (AOSIS) also put forward the use of insurance to address loss and damage (Roberts & Huq, 2015).[2] Since then, insurance has repeatedly formed part of the loss and damage debate,[3] most recently culminating in its inclusion in the WIM's mandate, illustrated by the indirect reference made to risk transfer and risk-sharing in the decision establishing the WIM. The WIM then incorporated insurance in its 2-year initial work plan and again in the subsequent 5-year rolling work plan. The latter includes comprehensive risk management approaches in strategic work stream C, where finding climate risk solutions through insurance is identified as a priority activity for 2019–2021 (WIM ExCom, 2017). In addition, work stream E, in particular sub-section 1(a)-(c), focuses on securing financial instruments to address loss and damage. Further support for wider coverage provided by insurance is evident in the mandate given to the Executive Committee of the WIM by the COP to develop a clearing house for risk transfer, the Fiji Clearing House for Risk Transfer, which was launched in 2018 (UNFCCC, 2016, 2018).[4]

Insurance is a type of *risk transfer* that can be used to shift the risk of loss and damage from one entity to another in exchange for a premium.[5] One form of *risk transfer* is *risk pooling*, whereby risk can be aggregated if organized in a pool (regionally or nationally), which allows for premiums to be lowered as risk is spread both across multiple actors (Gewirtzman et al., 2018) and in geographical terms (UNFCCC, 2012). These insurance pools or schemes can exist on three different levels: micro-, meso-[6] and macro. A summary of insurance mechanisms in operation shows that most are established on a micro- or macro-level, which is consequently where this section will focus. *Micro-insurances* are implemented at a local level for low-income populations and are suitable to insure crops or livestock. Here, individuals create a pool of policyholders and the payouts are made directly to the individuals within the risk pool (Schäfer et al., 2016). In *macro-insurance* schemes the policyholder is a national government within an insurance pool consisting of other countries in a specific geographical region. Payouts in macro-insurance schemes are made to the respective governments who can then invest in rehabilitation measures.

Regardless of the level they operate at, insurance schemes fall under one of two types: *indemnity-based* or *index-based*. *Indemnity-based* insurance schemes evaluate the loss and damage after an extreme event, once a claim has been handed in, and make payouts based on this assessment. However, the assessments can result in long delays before money is dispersed. On the other hand, *index-based* insurance works on the basis of pre-determined parametric triggers for natural disasters, such as rainfall amount or wind speed. Once triggered, a payout is made, which results in quick relief payment as no post-disaster assessment is required. This is a major benefit of *index-based* insurance.[7] Due to its particular relevance to loss and damage in the context of the climate change regime, this articles deals exclusively with *index-based* insurance.

When looking at existing insurance schemes and insured and uninsured losses over several years, the numbers are striking. In Africa, losses from hydrological, climatological and meteorological events between 2013–2015 equated to USD 11.5 billion, of which only 810 million were insured (NatCatSERVICE (MunichRE), 2019a). Similarly, losses from tropical cyclones in the same years in the Caribbean were USD 101 billion of which only USD 44 billion were insured (NatCatSERVICE (MunichRE), 2019b). It is worth noting that MunichRE only counts major catastrophic cyclones that meet certain parameters,[8] and therefore the uninsured losses are likely to be much higher in relation to overall losses.

In the last decade, macro-insurance schemes through risk transfer have been a popular solution at the regional level. Although only four regional insurance schemes exist, three of them cover reasonably sized and vulnerable geographical areas, namely the Caribbean, the Pacific, and parts of Africa. The first successful regional insurance scheme, which has been in operation since 2007, is the Caribbean Catastrophe Risk Insurance Facility (CCRIF). It aims to provide index-based insurance against extreme weather events for Caribbean governments and had, by 2017, made an accumulated payout of USD 100 million to its members (SPC, 2017). In the Pacific, the Catastrophe Risk Assessment and Financing Initiative (PCRFI) insurance scheme is operative.[9] The scheme was effective in 2013 when it made its first payout of USD 1.27 million to Tonga ten days after Cyclone Ian hit the island (Bank, 2013, 2014).[10] However, the initiative's impact is questionable, as illustrated by a different example, namely the payout to Vanuatu in 2015 after Cyclone Pam. In this instance, the insurance covered USD 1.9 million against a total damage cost of USD 449.4 million, which equates to less than half a per cent of incurred cost (Government of Vanuatu, 2015).

Their African counterpart is the African Risk Capacity (ARC), where prospective member countries must provide a spending and allocation plan in the case of a payout before entering the pool. Under the ARC, members are scheduled to receive a payout within three to four weeks of the end of the rainfall season (Capacity, 2018). However the mechanism proved problematic when it mishandled Malawi's extended drought in 2015, which included major crop failure. The event was exacerbated by its coupling with a rare flood just before the drought, which resulted in major food shortages in subsequent years (Richards & Schalatek, 2017). The ARC's models initially assessed that the drought did not trigger the policy and therefore refused to make a payout. In light of the severity of the situation, the ARC re-assessed their model to reflect more realistic growing times for crops, and the policy was triggered (Richards & Schalatek, 2017). This example uncovers three major problems with climate insurance. First, the assessment process and correlated payout undermine the credibility of insurance because the parameters used to determine payouts were challenged and subsequently adjusted. Second, it illustrates that by not selecting the right or appropriate level of insurance, material damages are not covered as the sum of the payouts is insufficient. Third, while index-based insurance does not take into account the effects of interplay between climate hazards, in this case a devastating flood immediately followed by a period of extreme drought, it could be designed to do so. Extraordinary situations of this kind could potentially be overcome if regional insurance pools in developing countries were modelled after the European Union Solidarity Fund for natural disasters. In this scheme, payouts based on solidarity can be made to a country experiencing multiple disasters even if pre-set thresholds are not met (Hochrainer-Stigler, Linnerooth-Bayer, & Lorant, 2017).

Micro-insurance schemes are also being implemented in national and regional contexts. On a micro-scale, index-based insurance provides an incentive to adapt and reduce risk because the payment is based on the event and not the actual losses incurred. A farmer, for example, would therefore have an incentive to invest in more drought-resilient crops to keep losses to a minimum, as the payout amount remains the same regardless of good farming practices (Linnerooth-Bayer, Surminski, Bouwer, Noy, & Mechler, 2018). Also, good practices could potentially provide a standpoint for farmers to negotiate lower premiums. Reduced loss additionally means that more of the insurance payout can be spent on recovery. At the same time, it reduces or eliminates the moral hazard problem, whereby policyholders refrain from taking adaptive or risk-reducing measures as the insurance payout increases in conjunction with losses incurred. Micro-insurance can therefore influence individual business owners' choice of crops, provide security in times of uncertain crop returns and encourage independence from government or aid payments.[11]

An example of a regional micro-insurance scheme is the R4 Rural Resilience Initiative. This initiative is a risk pool involving the governments of Malawi, Ethiopia, Senegal, Kenya and Zambia, with Zimbabwe to follow, offering insurance against drought for agricultural livelihoods. A unique feature of this insurance is that farmers have the option of providing their labour for projects that reduce climate risk in the region, instead of paying a premium into the insurance pool (Blampied, 2016). In addition, *risk transfer* is complemented with other safety net provisions such as loans and savings. It is noteworthy, however, that the programme is highly subsidized by developed country governments and charities (Linnerooth-Bayer et al., 2018) and would not be able to operate independently. This example suggests that insurance is not always a sustainable financial solution for responding to loss and damage without being backstopped by reliable international financing.[12]

Moreover, the concept of insurance is either unknown or poorly understood in many developing countries (Richards & Schalatek, 2017, 2018). Insurance is primarily a Western concept and implementation can therefore be complex as it must be tailor-made to the cultural environment in which it operates (Baumgartner & Richards, 2019). People struggling to make ends meet tend not to see payment of an insurance premium as a priority (Baumgartner & Richards, 2019). In addition, the low uptake of insurance can be linked to levels of education and literacy, as well as the complexity of insurance policies coupled with a lack of trust in financial institutions and the government (Marr, Winkel, van Asseldonk, Lensink, & Bulte, 2016). Accordingly, insurance may in fact be unsuitable or inappropriate in many country contexts.

3. The unsuitability of current insurance schemes

Although the move towards insurance schemes as a response to loss and damage has received significant support from a variety of stakeholders (Schäfer et al., 2016), there are several problems associated with relying on insurance in this context, as mentioned above and discussed further below.[13]

3.1. Sudden and slow onset events, and non-economic loss and damage (NELD)

In addition to the challenges presented above, insurance schemes have been criticised for their unsuitability in addressing the increasing frequency and severity of sudden onset events, slow onset events, and NELD. Traditionally, insurance schemes have required insurable events to be sudden, unpredictable, and infrequent (Gewirtzman et al., 2018). However, climate change will lead to more frequent and more severe sudden onset events, as well as slow onset events like sea level rise and desertification (Hoegh-Guldberg et al., 2018). This change presents a challenge to the usefulness of insurance schemes since insurability criteria can no longer be met; the unpredictability criterion, for one, will become increasingly difficult to satisfy due to improved ability to predict weather-related disasters (Gewirtzman et al., 2018).

The heightened risk posed by the increased frequency and severity of sudden onset events may lead to insurance providers withdrawing coverage entirely due to its untenability (Thomas & Leichenko, 2011). Given that the insurance industry is fundamentally designed for profit, climate change poses a significant challenge. Climate change essentially guarantees that loss and damage will be incurred and that payouts will need to be made for increasing amounts as impacts intensify. As such, if insurance is adopted as the main solution to address loss and damage, then individuals, communities, and countries may in fact be left without coverage should climate impacts become so frequent and severe that the insurance market is rendered unprofitable (Thomas & Benjamin, 2017). Alternatively, increased frequency and severity of events will likely lead to high premiums, making them unaffordable (Richards & Schalatek, 2017), and similarly resulting in limited insurance coverage. This leaves policy-holders in a highly vulnerable position and indicates that insurance alone may not be an adequate or comprehensive solution.

Many losses and damages will be the result of high-certainty, slow onset impacts (UNFCCC, 2012). It is widely recognized, both by scholars and the WIM Executive Committee, that insurance is unsuitable in the context of slow onset impacts (Blampied, 2016; Durand et al., 2016; Kehinde, 2014; Roberts et al., 2017; Schäfer et al., 2016; UNFCCC, 2012; Warner et al., 2013). Slow onset events by definition fail to meet the suddenness and unpredictability criteria for insurability since they essentially guarantee an inevitable outcome and a resulting financial loss (Roberts et al., 2017). Moreover, the magnitude of the loss would ofen prohibit effective pricing (Kehinde, 2014). Insurance is also unsuitable to respond to sudden onset events that are a result of slow onset events. For example, insurance would not normally cover damages stemming from a severe storm surge (sudden onset event), caused by sea level rise (slow onset event) (Thomas & Benjamin, 2017; Thomas & Leichenko, 2011).

Insurance is also ill-equipped to address small-scale events that are considered cumulative 'wear and tear' but which are ultimately a result of the adverse impacts of climate change (Linnerooth-Bayer et al., 2018; Moftakhari, AghaKouchak, Sanders, & Matthew, 2017). These types of impacts typically fail to meet the unpredictability requirement. Repeated small-scale nuisance flooding, for example, as a result of sea level rise, would aggregate into prohibitively high insurance premiums as its frequency increases.

Regardless of the type of impact, NELD will take place alongside economic loss and damage. While no uniform definition of NELD exists, the UNFCCC has concluded that non-economic losses can generally be defined as items that 'are not commonly traded in markets' (UNFCCC, 2013). Non-economic losses can be the direct or indirect result of climate change impacts and can take place in relation to three areas: individuals, society, or the environment (UNFCCC, 2013). For example, losses of life and health relate to private individuals, whereas losses in territory and cultural heritage can be more widely seen as losses experienced by society (Fankhauser, Dietz, & Gradwell, 2014). The types of assets or subjects covered by NELD are inherently uninsurable since they lack economic value, and thus do not lend themselves to insurance solutions in their current form (Bond, 2016; Durand et al., 2016). Beyond these limitations, it can be argued that assets can have a dual or multiple value, i.e. economic and non-economic, and insurance fails to account for this dual quality. For example, a house is not only a building worth its market value, it is also a home filled with personal possessions and memories and, in many cases, represents an anchoring in a neighbourhood which contributes to a sense of community. The mechanisms used to address the loss of a house because of climate change impacts must therefore also account for these non-economic values in order to truly compensate for that loss (Thomas & Benjamin, 2019).

Insurance schemes could, potentially, be substantially redesigned to cover NELD. This could be done by coupling NELD with insured economic assets or by automatically compensating for NELD whenever a consequential climate-related event occurs. A scheme of this kind would likely need to have a parametric trigger, as indemnity-based insurance cannot apply to unquantifiable harm. However, such an approach would give license to the international community to accept the NELD endured by communities and individuals, which is to say, it would require NELD to be considered quantifiable and essentially eligible for compensation, despite being incommensurable[14] (Wrathall et al., 2015). NELD is highly context-specific since the attachment to, and value of, certain assets or qualities depend on a given culture and tradition (Tschakert et al., 2017) and, furthermore, the subjective value of NELD can vary between premium-holders with a shared cultural background. Therefore, even if financial compensation for NELD would be acceptable to affected individuals or communities, the payout necessary to compensate for the loss and damage experienced would vary greatly between beneficiaries. This raises the risk of an inequitable compensation system, where some receive greater compensation than others for the same or comparable material loss. Conversely, if the same financial compensation were given for NELD regardless of local context, the compensation risks being inadequate in a multitude of other ways, thereby also making uniform payouts inappropriate. The problems of NELD are compounded when other relevant analytical perspectives are taken into account. Five variables are explored below.

3.2. Insurance schemes and common but differentiated responsibilities and respective capabilities (CBDR-RC)

The principle of CBDR-RC has its origins in Principle 7 of the Rio Declaration (UNFCCC, 1992), which acknowledges that 'in view of the different contributions to global environmental degradation, States have common but differentiated responsibilities'. In the climate change regime, CBDR-RC was included in the UNFCCC as a guiding principle (UNFCCC, 1992), and has played an essential role in the normative development of the regime, even if different parties have varying interpretations of its breadth and reach (Rajamani, 2014). A new qualification to the principle was added in the Paris Agreement, namely CBDR-RC 'in light of different national circumstances'. CBDR-RC can be operationalized in a number of ways, including through finance. Indeed, Richards and Schalatek (2017) argue that loss and damage forms part of the greater UNFCCC framework and, accordingly, CBDR-RC is applicable in its financing structures. In light of their limited historical responsibility and ability to pay for loss and damage, developing countries, particularly least developed countries, should not be burdened with financing incurred loss and damage alone (Adelman, 2016). Insurance raises issues in this regard, which will be explored below.

Given that the Paris Agreement explicitly excludes liability and compensation as part of loss and damage, by virtue of paragraph 51 of decision 1/CP.21 which limits the scope of Article 8, a clear benefit of insurance solutions is that they do not require the assignment of responsibility for incurred loss and damage (Lees, 2017). This is likely a key reason for the wide acceptance of insurance as a financial solution. Since this approach circumvents the need for specific events to be linked to anthropogenic climate change (Verheyen, 2015), the use of insurance schemes may allow for an expeditious process in securing a mechanism to distribute payouts.

The avoidance of the assignment of responsibility, while convenient, is problematic if seen in the context of the CBDR-RC principle. Applying CBDR-RC to loss and damage responses should, in theory, require some recognition of the disparity between states contributing to loss and damage and their capacity to address it. Insurance (without backstopping from international financial sources) fails to account for this necessity, as it places the burden for financing climate loss and damage on developing countries by requiring them to pay for insurance premiums, without consideration for the degree to which individual states have contributed to emissions (Page & Heyward, 2017; Voigt, 2008; Wallimann-Helmer, 2015). However, if insurance schemes were subsidized by developed countries, the demands of CBDR-RC would be satisfied to a greater extent.

Due to their *ex post* nature, insurance schemes are ill-suited to prevent or mitigate harm (Mahul et al., 2011). To mitigate harm and work in an *ex ante* manner, risk reduction and adaptation measures aimed at reducing vulnerability would need to be incentivised, such as the practice of the ARC. This could be achieved through setting premiums according to risk level, and encouraging policy-holders to reduce risk, with those at a higher risk of loss and damage paying higher premiums (Estrin & Vern Tan, 2016). While those facing risks of loss and damage are unable to control the primary risk, namely the negative impacts of climate change, they may be able to reduce their vulnerability to such impacts through adaptation or risk reduction measures, and thereby secure cheaper premiums.

Nevertheless, this approach has significant drawbacks. Some states are at higher risk of incurring loss and damage due to their geographical circumstances, such as having low land elevation and/or small land mass. Applying a risk-pricing mechanism for macro-insurance would therefore result in climate-vulnerable states paying higher premiums, despite the fact that these states are generally least able to afford them, and have also contributed less to current and historic carbon emissions (Adelman, 2016; Richards & Schalatek, 2018). Higher premiums might effectively punish states that are at high risk of incurring loss and damage, which seems contrary to the CBDR-RC principle and the overall spirit of the climate regime.

3.3. Insurance schemes and intergenerational equity

Article 3.1. of the UNFCCC states that 'Parties should protect the climate system for the benefit of present and future generations of humankind, on the basis of equity' (UNFCCC, 1992). This principle of intergenerational equity is not new to international environmental law,[15] and while the normative content of the principle remains debated, at its core, the principle aims to distribute resources and burdens between existing and future generations (Dupuy & Viñuales, 2018). Applied to the climate change context, Page argues that this means that existing generations should not make future generations worse off by depleting non-renewable resources without compensation (Page, 1999).

Insurance schemes entirely fail to respond to the requirements of intergenerational equity. Insurance schemes are inherently post-disaster risk tools that do not have a provision to prevent loss and damage from happening, but are designed to financially address a fraction of the loss and damage incurred. As payouts generally only cover a small amount of the damage incurred, they are not adequate to restore the *status quo ante* and hence, leave future generations materially, socially, and spiritually worse off. There has been little to no consideration of ways to incorporate the principle of intergenerational equity into insurance schemes in practice, but it may very well be possible. For instance, an additional share of the insurance payout could be made into an intergenerational equity fund within the impacted state. The accumulated funds could then be used for compensation payments for future generations. The viability of such a setup would, however, require further research.

3.4. Insurance schemes and economic inequality

Insurance schemes, in particular risk-pricing mechanisms, also raise issues in relation to economic equality. Persons living in poverty are often the most vulnerable to loss and damage and, when it occurs, are in greatest need of assistance in order to recover from this harm. A risk-pricing mechanism intended to prevent moral hazard may intensify the additional financial stress on those who are in greatest need of protection from climate change harm, as their high risk would make them eligible to pay high premiums.

Since insurance schemes favour current ownership structures, they also perpetuate existing economic inequality. The most economically disadvantaged would benefit the least from insurance solutions because the poor generally own fewer and less valuable assets (Roberts et al., 2017). If assets are insured and subsequently lost or damaged as a result of a climate change impact, the payout for economically disadvantaged individuals would be lower. While this appears logical, it risks perpetuating poverty (Richards & Schalatek, 2017) instead of alleviating it. Even if the material loss for the more affluent is greater, the impact of lost assets would be higher for the poor. Additionally, payouts by insurance schemes are generally insufficient in sum to fully restitute the loss incurred following a triggering event (Broberg, 2019). While payouts can be timely and thus assist the immediate recovery of a community or country (Schäfer et al., 2016), the economic relief provided by insurance payouts often constitutes only a fraction of incurred economic harm following an extreme weather event (Gewirtzman et al., 2018). In the case of the Malawi drought, for example, the insurance payout only covered about 2% of the total humanitarian need (Richards & Schalatek, 2018). Insurance payouts could thus have a limited effect on long-term and comprehensive recovery (Richards & Schalatek, 2018).

3.5. Insurance schemes and gender inequality

There is a risk of insurance mechanisms perpetuating or exacerbating existing inequality or marginalization in a recipient location (Schäfer et al., 2016), including in respect to gender inequality. The Paris Agreement preamble calls on its parties to respect, promote, and consider their obligations in relation to gender equality and the empowerment of women when undertaking climate action (UNFCCC, 2015).[16]

In country contexts where gender inequality is prevalent, this situation will likely be perpetuated by insurance schemes as the payouts may not follow a gender-sensitive distribution model. Insurance set-ups tend to favour men (Baumgartner & Richards, 2019), despite women being disproportionally negatively affected by the impacts of climate change (Connell & Coelho, 2018; Nansen Initiative, 2015). Women are often poorer, have higher health risks and needs, and lack access to institutionalized financial services such as a simple bank account, which would be required to receive direct insurance payments (Baumgartner & Richards, 2019). In such circumstances, insurance risks reinforcing existing unequal societal structures.

An OECD Development Centre study shows that women own only 15% of land titles in non-OECD countries (OECD, 2012). Globally, practices of land and property ownership remain problematic from a gender equality perspective as 'women in half the countries in the world are unable to assert equal land and property rights' (Roush & Liu, 2019). These discriminatory practices can be exemplified by inheritance laws that favour male over female inheritance and create disparity in property ownership. This type of discrimination is widespread in non-OECD countries, where discriminatory inheritance laws or practice are found in 86 out of 121 countries (OECD, 2012). Thus, if insurance schemes for loss and damage are in place in these countries (with no additional measures), women incurring loss and damage would receive smaller payouts as they generally own less property and will incur a smaller material loss. In this way, insurance may perpetuate existing gender inequality in an era of growing climate impacts.

3.6. Insurance schemes and human mobility

Issues arise as to what type of assets are considered insurable by insurance providers and whether a national practice for recognizing ownership could be followed. For example, constructions on informal (squatter) settlements might not traditionally be considered insurable property. As individuals and communities relocate due to resource scarcity or land erosion, informal settlements are likely to emerge, particularly in countries with little land mass or territory. This type of human mobility[17] can already be observed in island states such as Tuvalu, where outer islanders have begun to relocate to the main island, and reside in squatter settlements along the coast (De Albuquerque, 2013). These settlements are highly vulnerable to the adverse impacts of climate change such as floods, but are considered uninsurable as persons living in such settlements do not tend to have ownership rights to the land or the constructions on the land. This creates a situation where only some are eligible for compensation for their loss and damage whereas the most economically disadvantaged and already displaced receive nothing.

Further issues arise in relation to the expected increase in climate change-related human mobility (IPCC, 2014). Addressing human mobility forms part of the WIM's mandate under workstream D, and accordingly loss and damage solutions should be sensitive to the needs imposed by human mobility. Insurability criteria are misaligned with these needs and the consequences of irregular migration, given their lack of sensitivity to loss and damage experienced by migrants, such as that in relation to squatter settlements. Human mobility and relocation (be it planned or spontaneous) due to slow onset events and their related costs are also not considered by traditional insurance products. Planned relocation projects away from coastal zones due to climate change-related sea level rise, such as the village of Vunidogoloa in Fiji (Charan, Kaur, & Singh, 2017) or Kiribati's purchase of land in Fiji (Office of the President: Republic of Kiribati, 2014), fall within the purview of the WIM's mandate and should therefore qualify as an instance of climate-related loss. Current macro-insurance schemes are not designed to include cost recovery of any type of human mobility, be it planned or ad hoc, post- or pre-relocation. Accordingly, insurance schemes may be entirely unsuitable to address human mobility needs.

4. Conclusion

As the adverse impacts of climate change intensify, increasing loss and damage will be incurred. Thus, there is a pressing need for loss and damage finance solutions. Here, insurance schemes present an appealing option, as they are commonly used to compensate for loss and damage after unpredictable and discrete events, i.e. sudden onset adverse impacts.

However, insurance remains an inherently profit-driven market, so if there is no prospect of profit, companies will likely begin withdrawing coverage. In fact, the suitability of insurance will diminish over time as sudden onset events become more severe and more frequent as a result of climate change. In addition, insurance schemes fail to adequately address NELD.

This paper has also illustrated that insurance, as currently conceived, is misaligned with the spirit of the climate change regime. In their current form, insurance schemes do not embody the principles of CBDR-RC or intergenerational equity. Moreover, insurance schemes may exacerbate existing inequalities in social-economic structures, particularly in relation to gender and economic status.

Loss and damage is inherently complex and, in light of the foregoing, the potential of insurance schemes as they currently exist seems to be exhausted. Pursuing insurance as the primary tool for addressing loss and damage may distract and derail the debate from focusing on issues like sourcing new and additional finance for addressing loss and damage and making that finance available to climate-vulnerable countries and populations. For these reasons, it is concluded that insurance schemes, in their current forms, ultimately do not adequately address climate change challenges. Therefore, besides other potential tools, innovative and novel approaches to insurance schemes need to be developed in order to be relevant, effective, and just solutions to loss and damage. Overall, insurance must form part of a more holistic approach to respond to loss and damage.

Notes

1. This support is illustrated by the explicit inclusion of 'insurance solutions' for loss and damage in Article 8.4. of the Paris Agreement.
2. It should be noted that the idea of using insurance schemes to address natural disaster risk from climate change impacts is not restricted to loss and damage. Risk transfer and insurance were incorporated into the international climate change regime as early as 2010, through the establishment of the Cancun Adaptation Framework. Furthermore, the Hyogo Framework fo

Action, under the United Nations Office for Disaster Risk Reduction, also promotes insurance as a way for developing states to better cope with the aftermath of natural disasters (Mahul et al., 2011).

3. In the time between 1991 and 2013 loss and damage came up at various COPs. The Bali Action Plan asks for enganced action on adaptation which included loss and damage. Loss and damage was placed under the Cancun Adaptation Framework in 2010 with the decision to develop a work program on loss and damage. At COP 18 in Doha, it was decided to establish institutaional arrangements to addresss loss and damage which culminated in the establishement of the WIM. For a more detailed time line see (Roberts & Huq, 2015).

4. The Fiji Clearing House is a repository of information for insurance and risk transfer, and aims to connect stakeholders with experts that provide advice on tailor-made insurance solutions, with the purpose of increasing the coverage of insurance for impacts from natural disasters in developing countries. For further information see, unfccc-clearinghouse.org.

5. Other types of insurance instruments include risk retention and risk financing, which are usually initiated by the national government developing countries.

6. In *meso-insurance* schemes, an organisation (e.g. an NGO) creates a link between the insurer and an individual policyholder. Payments are made to the organisation, which then distributes payouts to the insured individual. For more on this see Blampied (2016).

7. For further reference on the benefits of index-based insurance for loss and damage see Schäfer et al. 'Making Climate Risk Insurance work for the Most Vulnerable: Seven Guiding Principles' (2016) Munich Climate Insurance Initiative and, in this issue, Broberg M., 'Parametric Loss nad Damage Insurance Schemes as a Means to Enhance Climate Change Resilience in Developing Countries' (2019).

8. MunichRE publishes data based on the five costliest events per year. Accounted events in the models that were run with NatCatSERVICE have caused at least one fatality and/or produced normalised losses $\geq$ US$ 100k, 300k, 1m, or 3m (depending on the assigned World Bank income group of the affected country).

9. It should be noted that the PCRFI is further not exclusively targeted at climate change impacts but also includes volcano outbreaks/eruptions and earthquakes. Pacific Island countries are discussing the establishment of a new facility that only covers climate change impacts. For further information see https://www.forumsec.org/22018-femm-pacific-islands-climate-change-insurance-facility/.

10. PCRFI is further not exclusively targeted at climate change impacts but also includes volcano outbreaks and earthquakes, but a new facility is currently being developed solely focusing on climate change impacts.

11. However, some incentives found at the micro level could, in theory, be introduced at the meso and macro levels through a tailored distribution design.

12. For studies on potential sources of finance see Durand et al. (2016), Roberts et al. (2017) and Richards and Schalatek (2017).

13. This paper evaluates the suitability of insurance schemes through the six variables described below. However, this is not an exhaustive list. Insurance may also raise other issues relating to ethnicity and distinctions between urban and rural settings, among others. However, due to the country context-specificity of these topics, they fall outside the scope of this paper.

14. Incommensurability in this context means that assets cannot be compared with any others, and therefore cannot be substituted. For further discussion of this concept, see Barnett et al. (2016) and Wrathall et al. (2015, p. 274–294).

15. See International Court of Justice Advisory Opinion on the Legality of the Threat or Use of Nuclear Weapons, ICJ Reports 1996, p. 226 and Gabčikovo-Nagymaros Project (Hungary/Slovakia), Judgment, ICJ Reports 1997, p. 7.

16. Preambular statements do not impose legal obligations but do inform the interpretation of the provisions of a treaty, by virtue of Article 31 of the Vienna Convention on the Law of the Treaties (adopted 1969, entry into force 1980) 1155 UNTS 18232. Accordingly the implementation of Article 8 on loss and damage must be informed by the Paris Agreement preamble.

17. In the context of the WIM, human mobility encompasses migration, displacement and planned relocation. For further information see WIM workstream D, available here https://unfccc.int/sites/default/files/resource/docs/2017/sb/eng/01a01e.pdf.

Disclosure statement

No potential conflict of interest was reported by the authors.

ORCID

Linnéa Nordlander http://orcid.org/0000-0002-5186-0372
Melanie Pill http://orcid.org/0000-0002-7235-2094
Beatriz Martinez Romera http://orcid.org/0000-0003-3742-1311

References

Adelman, S. (2016). Climate justice, loss and damage and compensation for small island developing states. *Journal of Human Rights and the Environment, 7*(1), 32–53.

Bank, W. (2013). *Pacific catastrophe risk insurance pilot - from design to implementation-Some lessons learnt.* Retrieved from https://reliefweb.int/sites/reliefweb.int/files/resources/Pacific_Catastrophe_Risk_Insurance-Pilot_Report_14071528129.pdf

Bank, W. (2014). Tonga to receive US $1.27 million payout for cyclone response [Press release]. Retrieved from http://www.worldbank.org/en/news/press-release/2014/01/23/tonga-to-receive-payout-for-cyclone-response

Barnett, J., Tschakert, P., Head, L., & Adger, W. N. (2016). A science of loss. *Nature Climate Change, 6,* 976–978.

Baumgartner, L., & Richards, J.-A. (2019). *Insuring for a changing climate a review and reflection on CARE's experience with microinsurance.*

Blampied, C. (2016). *Weathering a risky climate: the role of insurance in reducing vulnerability to extreme weather.* Retrieved from https://www.results.org.uk/sites/default/files/files/Weathering20a20Risky%20Climate.pdf

Bond. (2016). *Equitable, effective and pro-poor climate risk insurance: The role of insurance in loss and damage.* London: Bond Development and Environment Group. Retrieved from https://www.bond.org.uk/sites/default/files/resource-documents/deg_climate_risk_insurance_august_2016.pdf

Broberg, M. (2019). Parametric loss and damage insurance schemes as a means to enhance climate change resilience in developing countries. *Climate Policy.*

Capacity, A. R. (2018). Impacts. Retrieved from http://www.africanriskcapacity.org/impact/

Caribbean Catastrophe Risk Insurance Facility SPC. (2017). CCRIF reaches US$100 million in milestone payouts. Retrieved from https://www.ccrif.org/news/ccrif-reaches-us100-million-milestone-payouts

Charan, D., Kaur, M., & Singh, P. (2017). Customary land and climate change induced relocation—a case study of Vunidogoloa village, Vanua Levu, Fiji. In W. Leal (Ed.), *Climate change adaptation in Pacific countries* (pp. 19–33). Cham: Springer.

Connell, J., & Coelho, S. (2018). Planned relocation in Asia and the Pacific. *Forced Migration Review, 59,* 46–49.

De Albuquerque, C. (2013). *Report of the special rapporteur on the human right to safe drinking water and sanitation, Catarina de Albuquerque: Mission to Tuvalu.* Retrieved from http://www.un.org/en/ga/search/view_doc.asp?symbol=A/HRC/24/44/Add.2

Dupuy, P.-M., & Viñuales, J. E. (2018). *International environmental law* (2nd ed.). New York: Cambridge University Press.

Durand, A., Hoffmeister, V., Weikmans, R., Gewirtzman, J., Natson, S., Huq, S., & Roberts, J. T. (2016). *Financing options for loss and damage: A review and roadmap (1860–0441).* Bonn. Retrieved from https://www.die-gdi.de/uploads/media/DP_21.2016.pdf

Estrin, D., & Vern Tan, S. (2016). *Thinking outside the boat about climate change loss and damage: Innovative insurance, financial and Institutional mechanisms to address climate harm beyond the limits of adaptation. International Workshop Report.* Ontario. Retrieved from https://www.cigionline.org/sites/default/files/workshop_washington_march2016.pdf

The Executive Committee of the Warsaw International Mechanism for Loss and Damage (WIM ExCom). (2017). Report of the Executive Committee of the Warsaw International Mechanism for Loss and Damage associated with Climate Change Impacts - Annex The five-year rolling workplan of the Executive Committee of the Warsaw International Mechanism for Loss and Damage associated with Climate Change Impacts. Retrieved from https://unfccc.int/sites/default/files/resource/docs/2017/sb/eng/01a01e.pd

Fankhauser, S., Dietz, S., & Gradwell, P. (2014). *Non-economic losses in the context of the UNFCCC work programme on loss and damage.*

Gewirtzman, J., Natson, S., Richards, J.-A., Hoffmeister, V., Durand, A., Weikmans, R., ... Roberts, J. T. (2018). Financing loss and damage: Reviewing options under the Warsaw international mechanism. *Climate Policy, 18*(8), 1–11.

Government of Vanuatu. (2015). *Post-disaster needs assessment tropical cyclone pam, March* 2015. Retrieved from https://reliefweb.int/sites/reliefweb.int/files/resources/vanuatu_pdna_cyclone_pam_2015.pdf

Hochrainer-Stigler, S., Linnerooth-Bayer, J., & Lorant, A. (2017). The European Union Solidarity Fund: An assessment of its recent reforms. *Mitigation and Adaptation Strategies for Global Change, 22*(4), 547–563. doi:10.1007/s11027-015-9687-3

Hoegh-Guldberg, O., Jacob, D., Taylor, M., Bindi, M., Brown, S., Camilloni, I., ... Zhou, G. (2018). Global warming of 1.5°C. In V. Masson-Delmotte, P. Zhai, H. O. Pörtner, D. Roberts, J. Skea, P. R. Shukla, A. Pirani, W. Moufouma-Okia, C. Péan, R. Pidcock, S. Connors, J. B. R. Matthews, Y. Chen, X. Zhou, M. I. Gomis, E. Lonnoy, T. Maycock, M. Tignor, & T. Waterfield (Eds.), *An IPCC Special Report on the impacts of global warming of 1.5°C above pre-industrial levels and related global greenhouse gas emission pathways, in the context of strengthening the global response to the threat of climate change, sustainable development, and efforts to eradicate poverty.*

IPCC. (2014). Summary for policymakers. In *Climate change 2014: Impacts,adaptation, and vulnerability. Part A: Global and Sectoral Aspects. Contribution of Working Group II to the Fifth assessment Report of the Intergovernmental Panel on climate change.*

Kehinde, B. (2014). Applicability of risk transfer tools to Manage loss and damage from slow-onset Climatic risks. *Procedia Economics and Finance, 18,* 7. doi:10.1016/S2212-5671(14)00994-0

Lees, E. (2017). Responsibility and liability for climate loss and damage after Paris. *Climate Policy, 17*(1), 12. doi:10.1080/14693062.2016.1197095

Linnerooth-Bayer, J., Surminski, S., Bouwer, L. M., Noy, I., & Mechler, R. (2018). Insurance as a response to loss and damage? In R. Mechler, L. M. Bouwer, S. Th, S. Surminski, & J. Linnerooth-Bayer (Eds.), *Loss and damage from climate change Concepts, Methods and policy options* (pp. 483–512). Cham: SpringerOpen.

Mahul, O., Boudreau, L., Lane, M., Beckwith, R., & White, E. (2011). *Innovation in disaster risk financing for developing countries: Public and private contributions.* Retrieved from http://documents.worldbank.org/curated/en/609591468189559108/pdf/97430-WP-Box391475B-PUBLIC-DRFI-WRC-Paper-FINAL-April11.pdf

Marr, A., Winkel, A., van Asseldonk, M., Lensink, R., & Bulte, E. (2016). Adoption and impact of index-insurance and credit for smallholder farmers in developing countries. *Agricultural Finance Review, 76*(1), 94–118. doi:10.1108/AFR-11-2015-0050

Moftakhari, H. R., AghaKouchak, A., Sanders, B. F., & Matthew, R. A. (2017). Cumulative hazard: The case of nuisance flooding. *Earth's Future, 5*(2), 214–223. doi:10.1002/2016EF000494

Nansen Initiative. (2015). Disaster-induced cross-border displacement-Agenda for the protection of cross-border displaced persons in the context of disasters and climate change VOLUME 1. Retrieved from https://disasterdisplacement.org/wp-content/uploads/2014/08/EN_Protection_Agenda_Volume_I_-low_res.pdf

NatCatSERVICE (MunichRE). (2019a). *Relevant weather-related loss events in Africa 2013–2018*.

NatCatSERVICE (MunichRE). (2019b). *Tropical cyclone events in the Caribbean 2013–2018*.

OECD. (2012). 2012 *SIGI: Social institutions and gender index: Understanding the drivers of gender inequality*. Retrieved from https://www.oecd.org/dev/50288699.pdf

Office of the President: Republic of Kiribati. (2014). Kiribati buys a piece of Fiji [Press release]. Retrieved from http://www.climate.gov.ki/2014/05/30/kiribati-buys-a-piece-of-fiji/

Page, E. A. (1999). Intergenerational justice and climate change. *Political Studies, 47*(1), 53–66.

Page, E. A., & Heyward, C. (2017). Compensating for climate change loss and damage. *Political Studies, 65*(2), 356–372.

Rajamani, L. (2014). *Legal principles relating to climate change. Report of the international law association's committee on legal principles relating to climate change*.

Richards, J.-A., & Schalatek, L. (2017). *Financing loss and damage: A look at governance and implementation options*. Washington, DC. Retrieved from https://www.boell.de/sites/default/files/loss_and_damage_finance_paper_update_16_may_2017.pdf

Richards, J.-A., & Schalatek, L. (2018). *Not a silver bullet*. Washington, DC. Retrieved from https://us.boell.org/sites/default/files/not_a_silver_bullet_1.pdf

Roberts, E., & Huq, S. (2015). Coming full circle: The history of loss and damage under the UNFCCC. *International Journal of Global Warming, 8*(2), 141–157.

Roberts, J. T., Natson, S., Hoffmeister, V., Durand, A., Weikmans, R., Gewirtzman, J., & Huq, S. (2017). How will we pay for loss and damage? *Ethics, Policy & Environment, 20*(2), 208–226.

Roush, T., & Liu, A. S. (2019). Women in half the world still denied land, property rights despite laws [Press release]. Retrieved from https://www.worldbank.org/en/news/press-release/2019/03/25/women-in-half-the-world-still-denied-land-property-rights-despite-laws

Schäfer, L., Waters, E., Kreft, S., & Zissener, M. (2016). *Making climate risk insurance work for the most vulnerable: Seven guiding principles*. Bonn. Retrieved from http://www.climate-insurance.org/fileadmin/mcii/documents/MCII_PolicyReport2016_Making_CRI_Work_for_the_Most_Vulnerable_7GuidingPrinciples.pdf

Thomas, A., & Benjamin, L. (2017). Management of loss and damage in small island developing states: Implications for a 1.5°C or warmer world. *Regional Environmental Change, 18*(8), 1–10.

Thomas, A., & Benjamin, L. (2019). Non-economic loss and damage: Lessons from displacement in the Caribbean. *Climate Policy*.

Thomas, A., & Leichenko, R. (2011). Adaptation through insurance: Lessons from the NFIP. *International Journal of Climate Change Strategies and Management, 3*(3), 250–263.

Tschakert, P., Barnett, J., Ellis, N., Lawrence, C., Tuana, N., New, M., … Pannell, D. (2017). Climate change and loss, as if people mattered: Values, places, and experiences. *Wiley Interdisciplinary Reviews: Climate Change, 8*(5), 19. doi:10.1002/wcc.476

UNFCCC. (1992). United Nations Framework Convention on Climate Change. Retrieved from https://unfccc.int/resource/docs/convkp/conveng.pdf

UNFCCC. (2012). *A literature review on the topics in the context of thematic area 2 of the work programme on loss and damage: a range of approaches to address loss and damage associated with the adverse effects of climate change*. Retrieved from https://unfccc.int/resource/docs/2012/sbi/eng/inf14.pdf

UNFCCC. (2013). *Non-economic losses in the context of the work programme on loss and damage: Technical paper*. Retrieved from https://unfccc.int/resource/docs/2013/tp/02.pdf

UNFCCC. (2015). *The Paris agreement*. Retrieved from https://unfccc.int/sites/default/files/english_paris_agreement.pdf

UNFCCC. (2016). *Report of the conference of the Parties on its twenty-first session*. Retrieved from https://unfccc.int/resource/docs/2015/cop21/eng/10a01.pdf#page=2

UNFCCC. (2018). Fiji clearing house for risk transfer. Retrieved from http://unfccc-clearinghouse.org/about-us

Verheyen, R. (2015). Loss and damage due to climate change: Attribution and causation - where climate science and law meet. *International Journal of Global Warming, 8*(2), 158–169.

Voigt, C. (2008). State responsibility for climate change damages. *Nordic Journal of International Law, 77*(1–2), 1–22.

Wallimann-Helmer, I. (2015). Justice for climate loss and damage. *Climatic Change, 133*(3), 469–480.

Warner, K., Yuzva, K., Zissener, M., Gille, S., Voss, J., & Wanczeck, S. (2013). *Innovative insurance solutions for climate change: How to integrate climate risk insurance into a comprehensive climate risk management approach*. Bonn. Retrieved from http://collections.unu.edu/eserv/unu:1850/pdf11484.pdf

Whalley, A. (2016). *Ensuring climate risk insurance works for the poor*.

Wrathall, D. J., Oliver-Smith, A., Fekete, A., Gencer, E., Lepana Reyes, M., & Sakdapolrak, P. (2015). Problematising loss and damage. *International Journal of Global Warming, 8*(2), 20.

🔓 OPEN ACCESS

Parametric loss and damage insurance schemes as a means to enhance climate change resilience in developing countries

Morten Broberg

ABSTRACT

Article 8 of the 2015 Paris Agreement calls on the parties to cooperate to address 'loss and damage' associated with the adverse effects of climate change. Insurance is one of the solutions that the provision points to in this respect. This paper examines whether so-called parametric risk pooling may provide a useful tool to this end by examining lessons from three multi-country parametric risk pooling schemes. The examination shows that parametric risk pooling schemes require careful design in order to work optimally and that even then, they do not represent a miracle solution. In particular, the paper points out that the three schemes only provide a limited response in the period directly after a devastating hazard has struck. Therefore, the ensuing support required must come from other sources. The paper concludes that parametric risk pooling may provide a valuable tool for addressing loss and damage, but must be complemented by a range of other tools to comprehensively address the loss and damage challenges posed by climate change.

Key policy insights
- Parametric insurance schemes can provide a useful tool to respond to natural hazards caused by climate change in developing countries.
- Parametric insurance schemes require careful design and only provide interim coverage between the time when a natural hazard occurs and when classic humanitarian aid arrives.
- Due to continued financial limitations on the part of the insurance-takers, parametric insurance in developing countries is dependent upon donor assistance.

1. Background

In December 2015, representatives from 196 countries came together in Paris at the 21st UN Climate Change Conference of the Parties (COP 21) to negotiate a legally binding text on combating climate change within the United Nations Framework Convention on Climate Change (UNFCCC). During the negotiations leading up to this Paris Agreement, one of the key challenges concerned the notion of 'loss and damage'; an explicit recognition of this notion could be interpreted to mean that the developed countries were legally obligated to cover losses endured by vulnerable developing countries as a consequence of climate change.

The resulting Article 8 of the Paris Agreement unambiguously recognizes 'the importance of averting, minimizing and addressing loss and damage associated with the adverse effects of climate change'. However, in the Paris Agreement's adopting decision (decision 1/CP.21), the parties also explicitly agreed 'that Article 8 … does not involve or provide a basis for any liability or compensation' (UNFCCC, 2016).

Even if Article 8 does not establish a legal basis for liability or compensation, it is not an empty provision Rather, Article 8.3 lays down that the parties to the Paris Agreement 'should enhance understanding, action

This is an Open Access article distributed under the terms of the Creative Commons Attribution-NonCommercial-NoDerivatives License (http://creativecommons.org licenses/by-nc-nd/4.0/), which permits non-commercial re-use, distribution, and reproduction in any medium, provided the original work is properly cited, and is no altered, transformed, or built upon in any way.

and support, including through the Warsaw International Mechanism, as appropriate, on a cooperative and facilitative basis with respect to loss and damage associated with the adverse effects of climate change', and Article 8.4 provides a non-exhaustive list of eight areas where such 'cooperation and facilitation' could take place. The sixth area cited is 'Risk insurance facilities, climate risk pooling and other insurance solutions' (UNFCCC, 2016).

The objective of this paper is to consider the possibility of using so-called parametric insurance to address 'loss and damage' as set out in the Paris Agreement. It essentially asks whether parametric insurance may provide a useful tool in this respect. To answer this question, we shall not only consider risk pooling in general terms, but shall also examine the workings of three existing natural hazards[1] multi-country parametric risk pooling schemes operating in developing countries:[2] the Caribbean Catastrophe Risk Insurance Facility (CCRIF), the African Risk Capacity (ARC), and the Pacific Catastrophe Risk Assessment and Financing Initiative (PCRAFI).[3] In what follows, we shall first briefly consider the advantages of risk pooling in general and parametric insurance in particular (section 2) as well as the history behind 'insurance' being listed as one of the tools for addressing loss and damage (section 3). Subsequently, we look into the workings of the three aforementioned parametric risk pooling schemes (section 4) before assessing the pros and cons of each (section 5). Finally, we give our view on whether parametric risk pooling should be considered by donors in developed countries as a useful tool when promoting resilience against climate change in developing countries (section 6).

2. Advantages of parametric insurance

2.1. Risk pooling

Risk pooling is the principal method employed in modern insurance because it entails 'insurance takers' transferring the risk of some negative future occurrence to a joint pool. Said occurrence must be unforeseen, infrequent and its consequences must be appreciable (if the risk materializes). For the individual or business who procures the insurance, the risk's materialization has the potential to be devastating, so by transferring the risk (against payment of a premium) they convert an unpredictable high-stakes risk into a predictable annual premium and thus protect themselves. For the provider of risk insurance, however, the situation is rather different. The insurance provider relies on the 'law of averages' (or the 'law of large numbers') according to which the average of a large number of independent, identically distributed, random variables tends to fall close to an expected value. Therefore, by adding more risks to an insurance pool, the insurance provider can expect a reduction in variation around the expected value for the average loss per insurance taker. It follows that an increase in the number of insurance takers will reduce the probability that a risk pool will fail. Moreover, by taking on the combined risk of a substantial number of future occurrences, the risks will be neither unforeseen nor infrequent for the insurance provider. On the contrary, by pooling the risks, it becomes possible (to some extent) to calculate the number that will actualize, the frequency with which this will happen and the consequences of the realized risks. This allows the insurance provider to calculate premiums so that they cover the expected costs and generate a profit.

In addition, by pooling country-specific catastrophe risk insurance policies and placing them on the international reinsurance market as a single portfolio, the risk pool member countries can take advantage of regional risk diversification benefits and economies of scale, thereby making it possible to achieve a significant reduction in their premium (World Bank, 2016).[4]

In developing countries, losses caused by natural hazards such as droughts, floods or hurricanes have traditionally been borne by local populations and governments whilst international organizations, such as the United Nations' World Food Programme (WFP), have acted as de facto last resort insurers (Syroka & Wilcox, 2006). International organizations and humanitarian NGOs essentially aim to address a portfolio of risks that are almost certain to occur; the uncertainty is not whether the risk will materialize, but crucially when, where and at what magnitude. These organizations rely on uncertain funding flows from donors and are faced with the additional challenge that the overall global need for funding is growing (United Nations General Assembly, 2018). It is important to add that national governments in countries affected by natural hazards also provide funding to address disasters, typically by reallocating funds in their national budgets from planned development activities to crisis response. Such reallocations inevitably slow down development activities and reduce overall national resilience. The burden imposed by natural hazards is reflected in the fact that the annual cost of disasters in small developing states has been found to be 1.8% of GDP on average in these states (International Monetary Fund [IMF], 2016).[5]

Well-functioning risk pooling creates a stable flow of funding for responding to natural hazards which both evens out, and increases the predictability of, costs for the pool's member countries. Risk pooling also means that disaster response is financed before a hazard strikes and not in an ad hoc way in its wake. When the costs incurred by natural hazards are spread out at a predictable rate (the insurance premium) over a number of years and not suddenly met by the reallocation of funds from the national budget during a time of crisis, public and private actors' ability to plan ahead and implement strategies is considerably enhanced. In other words, well-functioning risk pools 'think ahead' to protect both development and resilience gains, and improve the quality of government support to affected regions and their populations. Additionally, if the risk pooling scheme is constructed so that it allows for the disbursement of the pay-out immediately after the hazard has struck, risk pools may prevent households from employing adverse coping mechanisms such as consuming seed grain or slaughtering livestock (Cervigni & Morris, 2016); that is, risk-coping mechanisms that are very likely to have long-term adverse effects on communities and cause development backsliding. To illustrate this point, Venton, Fitzgibbon, Shitarek, Coulter, and Dooley (2012) found that in Kenya, an early response to hazards in grassland areas could reduce the cost of food aid by 50% and the value of animal losses by 25%, or more, depending on the intervention used.

A risk pooling scheme essentially puts a price tag on hazards. This implies a paradigm shift since, historically, neither individual countries nor international organizations have made efforts to quantify the cost of retaining risk at the national level. In developing countries, ex ante risk reduction activities are typically undertaken using funds from national budgets, whereas the cost of hazard response is often borne by the international community through humanitarian aid. This makes it difficult for countries to fully comprehend the value of risk reduction or the full cost of managing hazards. Putting a price on the cost of hazards allows countries to make informed decisions regarding climate adaptation, disaster risk reduction expenditures, and contingency plans. For example, if a country knows that the cost of responding to a drought in a particular area would be USD 20 million, and the corresponding (extra) cost of distributing drought resistant seed to the same area is USD 5 million, it has an objective basis on which to make a decision. This contrasts with classic donor led hazard responses which generally bypass national treasuries and thereby do not allow the affected countries administrations to carry out objective assessments of the cost of ex post hazard response versus other ways of addressing hazards.

It must be acknowledged, however, that hazard insurance is not an exact science and, furthermore, that the overt quantification of hazards may be used for political purposes meaning that hazard risk pooling can potentially be used to malicious ends; for example, in section 5.2 below, we shall show that a government may use the quantification to argue that living in a given hazard prone area has become unsustainable and on this basis may cut off public support to the area.

2.2. Parametric insurance

This paper is concerned with the pros and cons of parametric risk pooling which is a particular kind of insurance pooling. Parametric risk pooling differs from traditional insurance schemes since pay-outs are not based on an assessment of the actual post-event losses, but are instead triggered by certain pre-defined parameters being met. Thus, when entering into a parametric insurance policy, the parties to the insurance agreement define which categories of event trigger the right to payment as well as the size of these payments. For example, they may stipulate that a hurricane of a certain magnitude hitting one of a number of predetermined cities engenders the payment of a defined sum. These predefined events (in this case, the hurricane) are correlated to a parameter or to an index of parameters (e.g. the magnitude of the event and the cities hit). Parametric insurance means that when an insurance event occurs, it is relatively straightforward to establish objectively whether the conditions for payment have been fulfilled and, if they have, the size of payments due. Consequently payment can be made swiftly as there is no delay from the evaluative process of calculating actual losses. Countries impacted by a hazard therefore often receive the funding they urgently need shortly after being hit, allowing them to address challenges almost immediately, which is important since speed is often crucial in the aftermath of these types of hazards.

The speed of parametric risk pooling systems nevertheless comes at a cost: the parametric coverage is modelled against expected consequences (economic 'losses' or 'response costs') of a specific hazard, so if the chosen parameters do not duly reflect the actual post-hazard consequences, the pay-outs may not match actual losses/response costs.

In this paper we shall examine three existing multi-country parametric risk pooling schemes in developing countries that were implemented to mitigate the risks caused by certain natural hazards. The natural hazards covered by these three parametric insurance schemes are infrequent, devastating events for which we cannot (with any degree of certainty) know the timings, locations and severities beforehand. But, based on historical data, it is possible to make reasonably accurate predictions about their frequency and magnitude. As we shall see in section 4.2, this information helps entities to tailor the coverage of individual policies, thereby ensuring that the insurance terms are triggered only in situations where the countries may be expected to suffer significant losses and/or incur response costs due to the hazard that would be difficult for them to shoulder alone. The schemes are therefore tailored so that hazards that are expected not to cause significant losses will not be covered – simply because the intention is to cover only the most significant events. A telling example of the consequences of this tailoring occurred in 2013 when the Solomon Islands suffered a magnitude 8.0 earthquake which caused substantial losses across the islands. Although the Solomon Islands had taken out parametric earthquake insurance with the PCRAFI, the losses were suffered far from the economic centre of the archipelago, so the impact (particularly on core government services) could not be classified as significant according to the parametric insurance policy, which stipulated that pay-outs would only be triggered if the earthquake had a certain seismic impact in the country's economic centre. Since the parametric threshold was not reached, the earthquake did not qualify to receive payment under the PCRAFI insurance scheme (World Bank, 2015).

A parametric insurance scheme relies on catastrophe risk models based on robust datasets. These models are used to assess the risk, but are also relevant to determine the triggering event. For example, if parametric insurance is taken out against the effects of a hurricane, the model may include criteria such as wind speed and direction in a specific location, and the trigger could well be wind speed in this specific location. In order to guarantee an unambiguous and independent determination of whether the trigger has been met, our three risk pooling schemes rely on data from internationally recognized bodies; the CCRIF, for example, relies on earthquake coverage data from the United States Geological Survey. In addition, an independent third party is designated to calculate pay-outs.[7]

3. The historic link between insurance and loss and damage

The link between insurance and the international fight against the adverse consequences of climate change is far from new. In order to understand the role which insurance plays in the context of loss and damage it is useful to look at how the concept of insurance has historically informed the development of the notion of 'loss and damage'. For instance, when the UNFCCC was being negotiated in 1990–1992, the then newly formed Alliance of Small Island States (AOSIS) proposed that '[t]here should be established as an integral part of the Framework Convention on Climate Change, … a separate International Insurance Pool to provide financial insurance against the consequences of sea level rise'. Furthermore, '[t]he resources of the Insurance Pool should be used to compensate the most vulnerable small island and low-lying coastal developing countries for loss and damage arising from sea level rise' (Wilford, 1994). Whereas the AOSIS proposal was not accepted, Article 4.8 of the UNFCCC explicitly listed insurance as one of the ways of meeting 'the specific needs and concerns of developing country Parties arising from the adverse effects of climate change and/or the impact of the implementation of response measures'.

In 1995, at COP 1, agreement was reached on a three-stage approach to adaptation funding. According to the decision of the COP, in the third and final stage '[m]easures to facilitate adequate adaptation, including insurance, and other adaptation measures' were envisaged for particularly vulnerable countries or regions (UNFCCC, 1995). In other words, in 1995, within the UNFCCC's legal framework insurance was strengthened as a tool to counter the consequences of climate change, albeit as part of climate change adaptation at this time.

In 2003, the COP held two workshops, targeted towards developing countries, focused on the implementation of insurance-related actions and their importance in addressing the effects of climate change. One of the recommendations that came out of the workshops was to consider establishing a network for international risk transfer via an international insurance pool aimed primarily at property losses due to climate change in poor communities in the least developed countries (Linnerooth-Bayer, Mace, & Verheyen, 2003; UNFCCC, 2003). In subsequent regional follow-up workshops, the questions of insurance, risk management and risk reduction were raised anew and a number of tools relating to insurance were identified as having the potential to minimize loss and damage and concurrently create systems that could provide resources for tailored solutions towards rehabilitation in small island developing states (SIDS) (UNFCCC, 2007).

At COP 14 in 2008, the questions of risk sharing and risk transfer mechanisms (such as insurance) were brought up again, and the idea to introduce a loss and damage mechanism was put forward. Attention was particularly drawn to climate change impacts that cannot be prevented through mitigation and adaptation efforts, and to the fact that the existing regulation did not cover this 'residual loss and damage' (Mace & Verheyen, 2016). In 2010, at COP 16, as part of the Cancun Agreements, the COP went even further by recognizing 'the need to strengthen international cooperation and expertise in order to understand and reduce loss and damage associated with the adverse effects of climate change, including impacts related to extreme weather events and slow onset events' (UNFCCC, 2011). The COP therefore pledged to establish a work programme in order to consider approaches to addressing loss and damage associated with climate change in those developing countries that are most vulnerable to this phenomenon (UNFCCC, 2011). At COP 17 in 2011, the COP adopted a 'Work programme on loss and damage' (UNFCCC, 2012), and at COP 18 in 2012, decided that an institutional arrangement should be established during COP 19 'to address loss and damage associated with the impacts of climate change in developing countries that are particularly vulnerable to the adverse effects of climate change' (UNFCCC, 2013).

Thus, when the COP met in Warsaw in 2013 (COP 19), 'the Warsaw international mechanism for loss and damage, under the Cancun Adaptation Framework[8] ... ' was established ' ... to address loss and damage associated with impacts of climate change, including extreme events and slow onset events, in developing countries that are particularly vulnerable to the adverse effects of climate change ... ' (UNFCCC, 2013). The decision establishing the Warsaw International Mechanism does not explicitly refer to insurance as a means to address loss and damage. However, reference is provided indirectly since in paragraph 5(c) it refers to Decision 3/CP.18, paragraph 6, which in its letter (b) lists 'risk reduction, and risk transfer and risk-sharing mechanisms', i.e. tools that also cover risk pooling (UNFCCC, 2013).

Therefore, when the COP adopted the Paris Agreement in 2015, Article 8's introduction of loss and damage can be viewed as the outcome of more than two decades of development since the first proposal in 1991. Two observations must be made. Firstly, the provision's legislative history not only shows that loss and damage is aimed at residual losses that cannot be addressed by mitigation and adaptation, but also that a driving force behind the recognition of loss and damage is to ensure financial support from wealthy countries for those that are most vulnerable. Secondly, insurance, and therefore risk pooling, has been a key aspect of loss and damage from the very beginning (Gewirtzman et al., 2018).

4. Workings of existing developing country multi-country parametric risk pooling schemes

4.1. Existing multi-country parametric risk pooling schemes

This paper is concerned with the three existing natural hazards multi-country parametric risk pooling schemes operating in developing countries. The first of the three schemes to come into existence – the CCRIF – was an initiative of the Caribbean Community (CARICOM) Heads of Government who approached the World Bank with a suggestion to create a multi-country sovereign risk pool. The scheme was set up on the basis of contributions to a multi-donor trust fund complemented by the membership fees paid by the member countries. The CCRIF began offering insurance coverage to Caribbean governments in the event of hurricanes and earthquakes in 2007, and against excess rainfall in 2013. By 2018, the CCRIF had made total payouts of about USD 139 million (CCRIF SPC, 2018).

In 2011, following the creation of the CCRIF, the World Bank also supported the establishment of the Pacific Catastrophe Risk Insurance Pilot, an integral part of the PCRAFI (World Bank, 2015).[9] This pilot scheme was largely driven by the World Bank in cooperation with several other donors. Based on the experience of the pilot scheme, in 2015 the Pacific countries decided to create the PCRAFI Facility which is set to run until 2021. This scheme is also heavily dependent upon international donors and by latest estimates (2018) has made cumulative pay-outs equivalent to USD 6.7 million (World Bank, 2018a, 2018b).

The ARC was developed as a joint project of the African Union (AU) and the United Nations World Food Programme (WFP). It became a specialized agency of the AU in 2012 and in 2018, counted 33 AU countries as members. Its insurance scheme is based on interest-free loans from donors rather than grants. With regard to operational costs, the ARC receives grant funding from donors and administrative support from the WFP. By 2018, the ARC had made cumulative pay-outs equivalent to more than USD 36 million ('ARC Impact', n.d.). As will be clear, all three risk pooling schemes were established with the assistance of international donor agencies and due to the continued financial limitations on the part of the insurance-taking countries, all three schemes still depend on financial and administrative support from the donors, and all three schemes continue to depend, in part, on external funding. This is unsurprising; before the introduction of multi-country risk pooling systems, their members would have relied almost entirely on international assistance when facing severe natural hazards, whereas now the member countries have recourse to schemes which, while still imperfect, are designed to prepare them to approach disaster in ways that are cost-effective and also, crucially, limit human suffering.

4.2. Tailoring the coverage to the individual member country

The three risk pooling schemes provide coverage tailored to individual member countries; each country can select which risks they are ready to shoulder themselves and which risks they would ask assistance for via the risk pool, as well as stipulating the amount of the pay-out in the event that the policy is triggered. Choices in risks and pay-outs have a direct impact on the premium to be paid and therefore member countries may opt for a premium that suits their finances by choosing a less generous coverage. The tailoring of each member country's policy is based on three parameters:

(i)   The attachment point (i.e. the severity of the event that gives rise to a payment, measured by statistical likelihood of occurrence in terms of years) (Clarke & Hill, 2013)
(ii) The exhaustion point (i.e. the severity of loss – related to the triggering event – at or above which point the maximum payment is triggered) (CCRIF, 2012)[10]
(iii) The amount (in US dollars) of the maximum pay-out

An insurance agreement that may be optimal from an economic point of view may, however, not be attractive to politicians. For example, the CCRIF offers coverage for natural events that may be expected to occur as frequently as every five years (a 'five-year return period'). For a politician, it might be interesting to have a pay-out under the insurance scheme, on average, every five years. However, to 'buy-into' pay-outs every five years means paying higher insurance premiums than, for instance, a 15-year pay-out would, and so the more frequent pay-outs may make risk pooling less efficient from an economic point of view. With regard to the ARC, their modus operandi is that claim payments should not be made 'to any country more frequently than once every five years, on average' and that, from a welfare perspective, it would be better still to reduce 'the claim payment frequency to once every eight or 10 years on average, and increasing the level of coverage for those extreme years' (Clarke & Hill, 2013).

4.3. First-response relief

The three schemes discussed in this paper are primarily intended to provide quick, efficient first-response relief and do not have the structure or resources required to manage all losses following an extreme event. First-response relief must therefore be followed by more long-term relief strategies such as loans and humanitarian aid. The schemes fill the, often crucial, void between occurrence of a hazard and the arrival of traditional assistance.

The schemes provide short-term financial liquidity via a quick infusion of capital that allows the member countries' governments to respond immediately to the needs of the affected communities. The schemes should not be viewed as a stand-alone solution but, on the contrary, as a tool to be complemented by a network of others.

5. Assessing the parametric risk pooling schemes

5.1. Economic aspects

Arguably, the principal advantage of the three schemes is that, by shouldering the risk together, each member country within a risk pool becomes less vulnerable to the immediate consequences of a climatic event.[11] In addition, by joining forces in a collective risk pool, the member countries can make significant economic savings on premiums (Ghesquiere & Mahul, 2012). For example, the ARC has estimated that 'by collectively pooling and diversifying their risks across the continent, countries save up to 50% in the cost of emergency contingency funds' (Beavogui, 2016).

As has been observed in section 2.2 above, parametric risk pooling schemes make it possible to provide immediate pay-outs to affected member countries once a hazard has struck, allowing them to respond very quickly to situations on the ground. For instance, the ARC observes that an

> analysis by the Boston Consulting Group shows the potential benefit of ARC outweighs the estimated cost of running ARC by 4.4 times compared to traditional emergency appeals for assistance, as a result of reduced response times and risk pooling. This means one dollar spent on early intervention through ARC saves four and a half dollars spent after a crisis is allowed to evolve. (ARC, 2016b)[12]

The view that a speedy pay-out from a risk pooling scheme in itself provides substantial benefits vis-à-vis more traditional (and slower) interventions has, however, been challenged. A cost–benefit analysis of the ARC in 2013 found that there were potential benefits from an early pay-out, as speeding up the disbursement of aid reduced the economic losses faced by households. However, they concurrently identified a number of conditions that must be met in order for these benefits to materialize. In particular, effective contingency plans are essential since without an appropriate distribution system the assistance may not reach vulnerable populations in a timely manner, even if ARC were able to make an early pay-out (Clarke & Hill, 2013).[6]

The picture becomes even more blurred since the United Nations Office for the Coordination of Humanitarian Affairs (OCHA) asserts that there is insufficient evidence to back the view that a swift reaction after the hazard has struck in combination with efficient contingency planning is significantly more cost-effective than a late reaction (OCHA, 2011). It is important to observe, however, that the OCHA does not deny that a swift reaction to natural disasters in combination with efficient contingency planning is probably significantly more cost-efficient than a late reaction; the organization merely points out that the stance is not evidence-based. In the author's opinion, an early response likely acts as a safety net which prevents those who are affected by the hazard from engaging in costly risk-coping mechanisms.

5.2. Broader societal adaptation to hazards

Insurance schemes can not only be used as a means to protect developing countries against the economic consequences of natural hazards, but also to promote societal adaptation to future hazards through education and partnerships. In addition, the ARC has taken an important step forward in requiring member countries to provide elaborate contingency plans before they can take out insurance. This broader, more holistic, approach to the implementation of risk pooling mirrors the overall ideas underlying Article 8 of the Paris Agreement on loss and damage.

The assurance of a pay-out in the event that hazards occur allows countries to plan response strategies ahead of time in a way that is not possible when hazard response funds are only obtained on an ad hoc basis after the hazard has struck (as is the case with humanitarian aid). Moreover, whereas the pay-outs from risk pools are channelled through the government system, this is often not the case with ad hoc funding. Channelling the pay-outs through government systems gives the government in a hazard-stricken country control over its own disaster response as well as providing better insight into the amount of hazard funding they receive.

In practice, policies within developing country risk pooling schemes put a 'value tag' on different geographic parts of member countries. For example, if a member country wishes to take out insurance cover against drought in a specific location where the likelihood of a severe and costly drought is substantial, the premium may be prohibitively high. In areas where this is the case, it may be reasonable to ask whether, from an economic point of view, the substantial risk of a devastating disaster makes local communities unsustainable. The answer to this query essentially comes down to politics. On the one hand, the pricing aspect of hazard risk insurance could give countries the tools to move from ex post responses to ex ante risk management since it accords governments, businesses and households the ability to plan before a hazard occurs and to agree on rules and processes for procuring, securing and spending relief funds (World Bank, 2017). On the other hand, because the scope of insurance coverage is, first and foremost, a political question, the high cost of hazard risk premiums may be used as an argument to label a given community unsustainable and therefore not eligible to receive government funding (and this refusal to invest in the community may go well beyond taking out hazard risk insurance). Thus, the geographic 'value tag' that hazard risk policies put on different parts of societies is a double-edged sword; it can be either a blessing or a curse for the affected communities.

5.3. Need for continuous updating and fine-tuning

As has been explained in section 2.2 above, parametric insurance is based on the premise that if certain pre-determined parameters are met, then certain losses can be expected. On this basis the insurance taker and the insurance provider agree on premiums and pay-outs. This approach has two apparent weaknesses. Firstly, the parameters may turn out to be misleading; for example, the losses that were only expected to occur when the parameters have been reached in fact occur below them. In this situation, the insurance taker suffers losses of a size that were intended to be covered by the insurance scheme, but receives no pay-out. Secondly, even if the parameters are an accurate reflection of the actual situation when the insurance policy is entered into, factual changes may occur so that when a hazard strikes, the parameters no longer reflect reality. A telling example of this occurred in 2016 when Malawi experienced a severe drought. The Malawian government had taken out drought insurance with the ARC and estimated that 6.5 million people would require food assistance. However, according to the ARC's parametric model, only approximately 20,000 people had been sufficiently adversely affected to trigger the parameters. In light of this huge discrepancy, the ARC undertook field research and discovered that farmers had switched to growing a different type of crop than the parametric model assumed. To correct for this, the ARC re-customized its software to generate an updated model that provided a better representation of the situation on the ground which led to an insurance pay-out to the Malawian government (ARC, 2016a; Reeves, 2017). This example shows that it is necessary to customize and constantly reevaluate the risk modelling software's parameters so that a member country's risk profile is accurately reflected and it can take years, with regular updating, to identify and integrate the correct data sets for a refined country customization (ARC & Oxford Policy Management, 2018).

6. Conclusion

To conclude on the question wether parametric risk pooling provides a useful tool to addres loss and damage, it may be useful to oberve that the history behind the introduction of the loss and damage provision in the Paris Agreement clearly shows recognition of the need for cooperation between developed country donors and developing countries vulnerable to climate change. In other words, parametric insurance schemes provide a solid platform for developed country donors to help vulnerable developing countries counteract some of the most destructive effects of climate change. This is reflected in the 2015 G7 InsuResilience Initiative led by the German government, which aims to provide 'access to direct or indirect insurance coverage against the impacts of climate change for up to 400 million of the most vulnerable people in developing countries by 2020' ('G7 Climate Risk Insurance Initiative', 2015). The objective of InsuResilience is to cover 180 million people by scaling up the African, Pacific and Caribbean risk pools (discussed in this paper), and through the publicly owned German KfW Development Bank's Climate Insurance Fund. In 2017, at COP23, the German government launched the InsuResilience Global Partnership to act as a multi-stakeholder community that ensures coordinated global action and improves access to knowledge regarding climate change (InsuResilience Global Partnership, 2018; UNFCCC, 2017). The results of the InsuResilience Global Partnership remain to be seen, but the initiative already provides a platform for developed country donors to help vulnerable developing countries to establish risk pooling schemes which can address the challenges posed by climate change.

Whereas the InsuResilience initiative may provide important momentum to both existing and new parametric multi-country risk pooling schemes in developing countries, the above examination of the three existing multi-country parametric risk pooling schemes has clearly shown that the schemes have both strengths and weaknesses. On the one hand, the examination has made it clear that, when carefully designed, the schemes may constitute a valuable tool to quickly and efficiently address some of those adverse consequences of climate change that are not avoided through mitigation and adaptation (Horton, 2018; Linnerooth-Bayer, Surminski, Bouwer, Noy, & Mechler, 2019). On the other hand, the examination has equally shown that parametric risk pooling schemes only provide a partial solution. The causes to this can be grouped into two broad categories. The first category covers the weaknesses that are inherent in the technical set-up of the parametric risk pooling schemes, namely the fact that it may be difficult to identify the relevant parameters with sufficient accuracy and that it may prove necessary to continuously evaluate the parameters – as illustrated in the case concerning Malawi (section 5.3 above). The second category groups the weaknesses that essentially are caused by limitations on the available funding – such as the fact that the schemes only address infrequent devastating events, and only finance the first, limited response after a hazard has struck.

Even though experience from the schemes examined in this paper shows that to some extent it is possible to address the weaknesses of parametric risk pooling schemes, it seems equally clear that it is impossible to fully remedy them. Thus, in a developing country context there are important limits to the accuracy of the data on which the parameters are based, and there are important restrictions on the available funding. Consequently, whereas it is true that parametric risk pooling can provide a valuable tool for addressing loss and damage in developing countries, our examination also shows it must be complemented by a range of other relief approaches in order to comprehensively address the loss and damage challenges posed by climate change.

Notes

1. The hazards covered by the schemes are: droughts (ARC), hurricanes (CCRIF, PCRAFI), earthquakes (CCRIF, PCRAFI), tsunamis (PCRAFI) and excess rainfall (CCRIF).
2. Since we are exclusively concerned with developing countries, we do not consider the South East Europe and Caucasus Catastrophe Risk Insurance Facility (Serbia and Macedonia).
3. For a detailed, critical examination and comparison of the three schemes, see Broberg and Hovani-Bue (2019).
4. The World Bank has estimated that the member countries of the Pacific Catastrophe Risk Assessment and Financing Initiative gained a 50% reduction in premiums as a result of the portfolio approach versus an individual country approach (World Bank 2016).
5. Moreover, both sudden and slow onset disasters caused by climate change have been found to place appreciably increased strains on the economies of countries in developing countries (Buhr et al., 2018).
6. An ARC payout must be able to reach affected households within 120 days (ARC, 2012).
7. The exact calculations of the pay-outs are laid down in the policies that have been underwritten.
8. The Cancun Adaptation Framework came out of COP16 which took place in 2010 in Cancun (Mexico).
9. The first transfer of risk to the international insurance market under the Pacific scheme took place in January 2013 (World Bank 2015).
10. In other words, the policy only covers losses up to a pre-specified limit. For example, in the 2009–2010 policy year Caribbean Catastrophe Risk Insurance Facility member countries selected exhaustion points equivalent to between 1-in75 and 1–in-200 year events (CCRIF, 2012).
11. It has, nevertheless, been suggested that insurance incurs substantial costs for the governments involved (Linnerooth-Bayer et al., 2019).
12. In addition to providing swift pay-outs, the risk pooling schemes also minimize the risk to investors and thereby support long term investments (ARC, 2016b)

Disclosure statement

No potential conflict of interest was reported by the author.

References

African Risk Capacity. (2012). *ARC response to the cost-benefit analysis of the African Risk Capacity*. African Risk Capacity. Retrieved from http://www.africanriskcapacity.org/wpcontent/uploads/2016/11/ARC_CBA_and_Response.pdf

African Risk Capacity. (2016a). *PRESS RELEASE – Malawi to receive USD 8M insurance payout to support drought-affected families – African Risk Capacity*. Retrieved from http://www.africanriskcapacity.org/2016/11/14/press-release-malawi-to-receive-usd-8m-insurance payout-to-support-drought-affected-families/

African Risk Capacity. (2016b). *The cost of drought in Africa*. African Risk Capacity. Retrieved from http://www.africanriskcapacity.org/wp-content/uploads/2016/11/arc_cost_of_drought_en.pdf

African Risk Capacity & Oxford Policy Management. (2018). *Building climate and disaster resilience in Africa: Lessons from the African Risk Capacity (ARC)*. Oxford: Oxford Policy Management. Retrieved from http://www.opml.co.uk/sites/default/files/Building-climate-and-disaster-resilience-in-Africa.pdf

Beavogui, M. (2016). *Case study: The African Risk Capacity (ARC) and Talking points*. London: African Risk Capacity.

Broberg, M., & Hovani-Bue, E. (2019). Disaster risk reduction through risk pooling: The case of hazard risk pooling schemes. In K. L. H. Samuel, M. Aronsson-Storrier, & K. Nakjavani Bookmiller (Eds.), *The Cambridge Handbook of disaster risk reduction and international law* (pp. 257–274). Cambridge: Cambridge University Press.

Buhr, B., Donovan, C., Kling, G., Lo, Y., Murinde, V., Pullin, N., & Volz, U. (2018). *Climate change and the cost of capital in developing countries - Assessing the impact of climate risks on sovereign borrowing costs*. UNEP Inquiry, Imperial College Business School, SOAS.

Caribbean Catastrophe Risk Insurance Facility. (2012). *Understanding CCRIF's hurricane and earthquake policies* (Technical Paper Series No. 1). The Caribbean Catastrophe Risk Insurance Facility. Retrieved from http://www.ccrif.org/sites/default/files/publications/TechnicalPaper1HurricaneEarthquakePoliciesAugust2012.pdf

Caribbean Catastrophe Risk Insurance Facility Segregated Portfolio Company. (2018). *Annual report 2017-2018*. Cayman Islands: The Caribbean Catastrophe Risk Insurance Facility. Retrieved from https://www.ccrif.org/sites/default/files/publications/CCRIF_Annual_Report_2017_2018_0.pdf

Cervigni, R., & Morris, M. L. (2016). *Confronting drought in Africa's drylands: Opportunities for enhancing resilience*. Washington, DC: World Bank : Agence Francaise de Development. Retrieved from http://uproxy.library.dc-uoit.ca/login?url=http://elibrary.worldbank.org/doi/book/10.1596/978-1-46480800-5

Clarke, D. J., & Hill, R. V. (2013). *Cost-benefit analysis of the African risk capacity facility*. International Food Policy Research Institute. Retrieved from http://www.ifpri.org/sites/default/files/publications/ifpridp01292.pdf

G7 Climate Risk Insurance Initiative: Stepping Up Protection for the Most Vulnerable. (2015). Retrieved from http://newsroom.unfccc.int/lpaa/resilience/g7-climate-risk-insurance-initiative-stepping-upprotection-for-the-most-vulnerable/

Gewirtzman, J., Natson, S., Richards, J.-A., Hoffmeister, V., Durand, A., Weikmans, R., ... Roberts, J. T. (2018). Financing loss and damage: Reviewing options under the Warsaw international mechanism. *Climate Policy*, 18(8), 1076–1086.

Ghesquiere, F., & Mahul, O. (2012). *Caribbean Catastrophe Risk Insurance Facility (CCRIF): disaster risk financing & insurance case study*. The World Bank. Retrieved from http://documents.worldbank.org/curated/en/319251467999348409/pdf/97469-BRI-Box391476B-PUBLICstudy-CCRIF-Final.pdf

Horton, J. B. (2018). Parametric insurance as an alternative to liability for compensating climate harms. *Carbon & Climate Law Review*, 12 (4), 285–296.

Impact – African Risk Capacity. (n.d.). Retrieved from http://www.africanriskcapacity.org/impact/

InsuResilience Global Partnership. (2018). *Final consultation document - concept note shaping the InsuResilience Global Partnership*. Retrieved from https://www.insuresilience.org/wpcontent/uploads/2018/05/20180502_Concept-Note_InsuResilience-Global-Partnership.pdf

International Monetary Fund. (2016). *Small states' resilience to natural disasters and climate change - role for the IMF* (IMF Policy Paper). Washington, DC: International Monetary Fund. Retrieved from https://www.imf.org/external/np/pp/eng/2016/110416.pdf

Linnerooth-Bayer, J., Mace, M. J., & Verheyen, R. (2003). *Insurance-related actions and risk assessment in the context of the UNFCCC* (Background Paper for UNFCCC Workshops).

Linnerooth-Bayer, J., Surminski, S., Bouwer, L. M., Noy, I., & Mechler, R. (2019). Insurance as a response to loss and damage? In R. Mechler, L. M. Bouwer, T. Schinko, S. Surminski, & J. Linnerooth-Bayer (Eds.), *Loss and damage from climate change: Concepts, Methods and policy Options* (pp. 483–512). Cham: Springer International Publishing.

Mace, M. J., & Verheyen, R. (2016). Loss, damage and responsibility after COP21: All options open for the Paris agreement. *Review of European, Comparative & International Environmental Law*, 25(2), 197–214.

Office for the Coordination of Humanitarian Affairs. (2011). *OCHA and slow-onset emergencies* (OCHA Occasional Policy Briefing Series No. 6). OCHA - Policy Development and Studies Branch (PDSB). Retrieved from https://www.unocha.org/sites/unocha/files/OCHA%20and%20Slow%20Onset%20Emergencies.pdf

Reeves, J. (2017). *The wrong model for resilience: How G7-backed drought insurance failed Malawi, and what we must learn from it*. ActionAid UK. Retrieved from http://www.actionaid.org/sites/files/actionaid/the_wrong_model_for_resilience_final_230517.pdf

Syroka, J., & Wilcox, R. (2006). Rethinking international disaster aid finance. *Journal of International Affairs, 59*(2), 197–214.

United Nations Framework Convention on Climate Change. (1995). *Decision 11/CP.1 initial guidance on policies, programme priorities and eligibility criteria to the operating entity or entities of the financial mechanism* (Document No. FCCC/CP/1995/7/Add.1). Berlin, Germany. Retrieved from https://unfccc.int/resource/docs/cop1/07a01.pdf

United Nations Framework Convention on Climate Change. (2003). *Implementation of article 4, paragraphs 8 and 9, of the convention progress on the implementation of article 4, paragraph 8 - Report on the UNFCCC workshops on insurance - Note by the Chair of the Subsidiary Body for Implementation* (Document FCCC/SBI/2003/11). United Nations Framework Convention on Climate Change. Retrieved from https://unfccc.int/resource/docs/2003/sbi/11.pdf

United Nations Framework Convention on Climate Change. (2007). *Vulnerability and adaptation to climate change in small island developing states - Background paper for the expert meeting on adaptation for small island developing States.* United Nations Framework Convention on Climate Change. Retrieved from http://unfccc.int/files/adaptation/adverse_effects_and_response_measures_art_48/application/pdf/200702_sids_adaptation_bg.pdf

United Nations Framework Convention on Climate Change. (2011). *Decision 1/CP.16 The Cancun agreements: Outcome of the work of the Ad hoc working group on long-term cooperative action under the convention* (Document FCCC/CP/2010/7/Add.1). Cancun, Mexico. Retrieved from https://unfccc.int/sites/default/files/resource/docs/2010/cop16/eng/07a01.pdf

United Nations Framework Convention on Climate Change. (2012). *Decision 7/CP.17 work programme on loss and damage* (Document FCCC/CP/2011/9/Add.2). Durban, South Africa. Retrieved from https://unfccc.int/sites/default/files/resource/docs/2011/cop17/eng/09a02.pdf

United Nations Framework Convention on Climate Change. (2013). *Decision 3/CP.18 approaches to address loss and damage associated with climate change impacts in developing countries that are particularly vulnerable to the adverse effects of climate change to enhance adaptive capacity* (Document FCCC/CP/2012/8/Add.1). Doha, Qatar. Retrieved from https://unfccc.int/sites/default/files/resource/docs/2012/cop18/eng/08a01.pdf

United Nations Framework Convention on Climate Change. (2016). *Decision 1/CP.21 adoption of the Paris agreement* (Document FCCC/CP/2015/10/Add.1). Paris, France. Retrieved from https://unfccc.int/resource/docs/2015/cop21/eng/10a01.pdf

United Nations Framework Convention on Climate Change. (2017). *Joint statement on the InsuResilience Global Partnership.* Bonn. Retrieved from https://www.bmz.de/de/zentrales_downloadarchiv/cop23/climate_risk_insurance/171205_Joint_Statement_Insu Resilience_Global_Partnership_COP23_Bonn.pdf

United Nations General Assembly. (2018). *Implementation of the Sendai framework for disaster risk reduction 2015–2030* (Report of the Secretary-General No. A/73/268). United Nations General Assembly. Retrieved from https://www.unisdr.org/files/resolutions/N1824255-en.pdf

Venton, C. C., Fitzgibbon, C., Shitarek, T., Coulter, L., & Dooley, O. (2012). *The economics of early response and disaster resilience: Lessons from Kenya and Ethiopia.* Department for International Development (DFID). Retrieved from https://www.gov.uk/government/uploads/system/uploads/attachment_data/file/67330/Econ-Ear-RecRes-Full-Report_20.pdf

Wilford, M. (1994). Insuring against sea level rise. In P. Hayes & K. Smith (Eds.), *The global greenhouse regime, who pays? Science, economics and North-South politics in the climate change convention* (Reprint.) (pp. 169–187). London: Earthscan.

World Bank. (2015). *Pacific catastrophe risk insurance pilot report: From design to implementation - some lessons learned.* Washington, DC: The World Bank. Retrieved from https://www.gfdrr.org/sites/default/files/publication/Pacific_Catastrophe_Risk_Insurance Pilot_Report_140715%281%29.pdf

World Bank. (2016). *Implementation completion and results report on a grant in the amount of US$ 4 million to the Pacific islands (Marshall Islands, independent state of Samoa, Solomon Islands, Kingdom of Tonga, and Vanuatu) for a Pacific catastrophe risk insurance pilot program* (No. ICR00003696). The World Bank. Retrieved from http://documents.worldbank.org/curated/en/655341475523018949/pdf/PICCatastropheRiskInsurancePilotSmallGrant-ICR-FINAL-09302016.pdf

World Bank. (2017). *Sovereign climate and disaster risk pooling - World Bank technical contribution to the G20.* Washington, DC: The World Bank.

World Bank. (2018a). *PCRAFI program – phase II: Furthering disaster risk finance in the Pacific.* The World Bank. Retrieved from https://financialprotectionforum.org/publication/4-pager-pcrafi-program-furtheringdisaster-risk-finance-in-the-pacific-2016-21

World Bank. (2018b). *PCRAFI program phase II: Furthering disaster risk finance in the Pacific – regional collaboration on climate and disaster risk financing.* The World Bank. Retrieved from https://financialprotectionforum.org/publication/brochure-pcrafi-program-phase-ii-furthering-disasterrisk-finance-in-the-pacific-2016-21

Non-economic loss and damage: lessons from displacement in the Caribbean

Adelle Thomas ⓘD and Lisa Benjamin ⓘD

ABSTRACT

Climate-induced displacement has direct implications for non-economic loss and damage, including threats to health and wellbeing and loss of culture and agency. Displacement due to extreme events is particularly challenging for small island developing states (SIDS) given their high exposure and vulnerability to tropical cyclones. Devastating hurricanes in the Caribbean in 2017 exposed non-economic loss and damage associated with prolonged displacement of entire island populations due to complete destruction of communities. Such was the case in Ragged Island, The Bahamas, where the entire population was displaced. This study assesses national policies, plans, legislation and reports from The Bahamas to determine non-economic loss and damage experienced by displaced residents and how the policy landscape addresses these issues. We find that non-economic loss and damage was acknowledged neither by the policy landscape nor by Government actions, but that there were likely health impacts and disruptions to sense of place and connection to the island. Failure to consider non-economic loss and damage also contributed to assessments that costs of rebuilding outweighed benefits. While existing literature has acknowledged policy deficiencies on loss and damage at the national level in SIDS, this study illustrates real-world impacts of these deficiencies. The case of Ragged Island highlights the need for SIDS to take the lead in developing national responses to loss and damage as they are currently experiencing severe impacts, which are intensified by the lack of clear policies, plans or strategies.

Key policy insights
- Climate-induced displacement is linked to non-economic loss and damage, particularly when displacement is prolonged and due to extreme events. However, these impacts are rarely addressed in national policies, plans or strategies.
- National policies that address climate-induced displacement and non-economic loss and damage are critical in order to reduce impacts experienced by displaced communities and expand feasibility assessments of rebuilding to include more than economic considerations.
- SIDS are currently experiencing non-economic loss and damage and should take the lead in developing national policies and strategies to assess, address and report on these impacts, particularly those SIDS that rely on international funders for recovery and response to extreme events.

1. Introduction

Loss and damage is a critical issue for small island developing states (SIDS) who are already experiencing impacts associated with climate change. The Alliance of Small Island States (AOSIS) was the first to raise the concept of loss and damage in the international policy arena at the United Nations Framework Convention on Climate Change (UNFCCC) and has been a strong advocate for advancing the issue, calling for provision of finance for incurred loss and damage and the development of robust mechanisms to provide support for projected impacts (Benjamin, Thomas, & Haynes, 2018). The majority of SIDS also included reference to loss and damage in their Nationally Determined Contributions (NDCs), emphasizing loss and damage that has already been experienced and stressing the importance of action by the UNFCCC to address the issue (Thomas & Benjamin, 2017).

For Caribbean SIDS, the 2017 hurricane season brought the issue of loss and damage to the forefront. Hurricanes Irma and Maria, both category 5 storms, passed through the region within two weeks of each other resulting in severe damages in many islands (Sullivan, 2017). Media and political attention focused on infrastructural damages and losses to industries such as agriculture and tourism along with their associated costs. The total economic damages of the storms are still unknown, with an estimated expense of over USD5.4 billion for only five of the countries affected (ECLAC, 2018).

These impacts from hurricanes whose severity was linked to climate change increased pressure from Caribbean SIDS at the 23rd UNFCCC Conference of Parties (COP 23) in 2017 for more progress to be made on loss and damage at the international level (Benjamin et al., 2018). One of the issues that these small islands advocated for was increased support for the UNFCCC Warsaw International Mechanism for Loss and Damage (WIM) and its Executive Committee (ExCom) in the hope that the body would be able to improve support and guidance on ways to address loss and damage at regional and national scales. SIDS also advocated for provision of finance for incurred loss and damage and establishment of the issue as an ongoing agenda item during the negotiation process (Benjamin et al., 2018).

However, while Caribbean SIDS have been vocal on loss and damage at international fora, at the national scale many do not have specific policies or mechanisms that address loss and damage in a holistic manner and few countries include components of loss and damage in other related policies, such as climate change adaptation or disaster risk reduction (Thomas & Benjamin, 2017). Discussion of loss and damage in the NDCs of these countries also places emphasis on economic costs stemming from infrastructural damage and sectoral losses associated with extreme events.

This focus on economic costs obscures the significant non-economic loss and damage – such as effects on health, sense of place and community cohesion – that was incurred from the 2017 hurricanes and the need for robust, national policies to address these impacts. In particular, approximately 3 million people from 16 countries and territories in the region were displaced due to the hurricanes including the unprecedented displacement of entire island populations in both Antigua and Barbuda and The Bahamas, resulting in two islands becoming completely uninhabited (IDMC, 2018). Displacement of residents due to extreme events- including hurricanes, floods and tropical storms- has long been an issue for Caribbean SIDS, although there has been little academic scholarship focused on this issue (Kaenzig & Piguet, 2014). Such displacement is a subset of the global internal movement of people due to extreme events, which in 2015 amounted to over 19 million individuals and is expected to continue to be a significant humanitarian challenge as climate change intensifies (United Nations Office for the Coordination of Humanitarian Affairs, 2017)

The non-economic loss and damage associated with the 2017 hurricane displacements, along with the effectiveness or otherwise of national approaches to address these movements, has not been subjected to in-depth attention or inquiry. This study addresses this gap by assessing the non-economic loss and damage associated with displacement of the entire population of Ragged Island, The Bahamas, as a result of Hurricane Irma. We also assess national policies, plans, legislation and documents submitted by The Bahamas to the UNFCCC including those focused on climate change, as well as policies on disaster risk reduction, to identify any included measures to address displacement and associated non-economic loss and damage. We highlight strengths and gaps in national policies, plans and legislation to address non-economic loss and damage components of displacement associated with extreme events and show how the experience of Ragged Island illustrates the challenges faced by many SIDS and offers lessons on how these policy deficiencies can have real world impacts in the small island context (Figure 1).

Non-economic loss and damage can be used as a lens to examine Government action and community responses post-disaster. Using such a lens, this study illustrates the trade-offs made by vulnerable communities between values and risks, and the importance of sense of place in decision making. Few studies assess patterns of displacement, its interaction with extreme events and the influence on non-economic loss and damage for SIDS. This study contributes to the existing evidence base and provides lessons for both national policies and

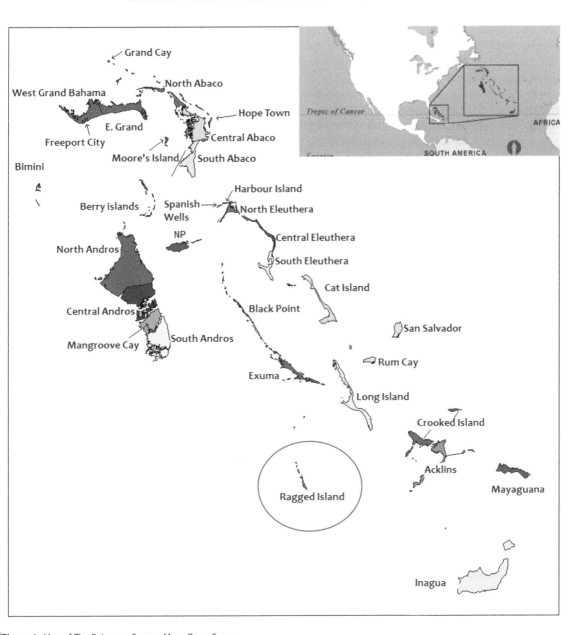

Figure 1. Map of The Bahamas. Source: Maps Open Source.

regional experiences, and further deepens understanding of the impact of non-economic loss and damage on Government and community action in the face of climate-induced loss and damage. While this paper focuses on the non-economic loss and damage experienced by the community in Ragged Island, it should be noted that significant economic loss and damage was also experienced by the community through loss of their homes, churches, schools, agricultural land, and infrastructure.

2. Non-economic loss and damage & displacement

The UNFCCC has described non-economic losses as those that are not commonly traded in markets, including loss of life, health, territory, traditional knowledge, culture, biodiversity, ecosystem services, as well as climate-induced displacement (UNFCCC, 2013). Non-economic losses can also include damage to, and loss of, physical properties such as traditional meeting houses, places of worship, artefacts, sacred places and communal land (Benjamin et al., 2018). These physical damages and losses can in turn cause damage to, or loss of, intangibles such as cultural values, traditions, senses of identity and loss of a sense of place. Non-economic losses can occur at a variety of spatial scales, and for many developing countries, these losses may be more significant than economic ones (UNFCCC, 2013). However despite their prevalence and importance, non-economic losses are rarely included in assessments of climate change impacts, and as a result are obscured and not addressed in risk analyses and policy-making (Serdeczny, Menke, & Thomas, 2018).

Climate-induced displacement is well-recognized as having the clearest connection to non-economic losses in comparison to other forms of human mobility, such as voluntary migration or planned relocation (UNFCCC, 2013). Displacement can leave individuals with a sense of being disconnected from themselves and from community members (Serdeczny, Waters, & Chan, 2016). Security, legal rights, dignity, social networks and agency are all threatened by displacement, particularly in instances where people are forced from their homes due to extreme hydrometeorological events (UNFCCC, 2013). Displaced households also face larger threats to health when placed in temporary housing, including greater rates of infant and child mortality (Chen, Halliday, & Fan, 2016).

Studies on displacement in SIDS due to both sudden and slow-onset events related to climate change are limited. Much research instead focuses on permanent migration of populations due to floods, tropical storms and sea level rise with particular attention paid to implications for international migration (Curtain & Dornan, 2019; Hill, 2016; Kaenzig & Piguet, 2014; Mortreux & Barnett, 2009). While impacts of natural hazards in SIDS has long been an area of inquiry, the focus has largely centered on implications for ecosystems, human mortality, infrastructure and economic activities (e.g. Cashman & Nagdee, 2017; Lewsey, Cid, & Kruse, 2004; Monioudi et al., 2018). There is also a long-recognized dearth of academic literature on experienced impacts of natural hazards in general, and on displacement due to extreme events in particular, for SIDS (Hamza, Koch, & Plewa, 2017; Méheux, Bank, Dominey-Howes, & Lloyd, 2007; Nurse et al., 2014; Thomas & Benjamin, 2018).

The few studies that do focus on displacement in SIDS due to extreme events have underscored the importance of the issue and have called for further research. Hamza et al. (2017) highlight that, although people living in SIDS are three times more likely to be displaced by disasters than people living elsewhere, there has been little analysis of the implications of such displacement due to the small total number of people that are affected in comparison to the effects experienced in larger countries. Their research indicates that the drivers of displacement are similar across SIDS in the Caribbean and the Pacific and include informal settlement development, and limited availability of land that is not highly exposed to disasters and poverty. Millar (2007) focuses on the lack of coordinated legal and administrative systems to prevent conflict and political instability spurred by displacement in SIDS and calls for the development of legal initiatives and policies to address these potential impacts. The majority of the limited information available on displacement in SIDS due to extreme events comes from government and relief agency reports and databases (e.g. CRED, 2019; Government of Dominica 2018). Other grey literature focuses on numbers of people displaced and does not include assessment of non-economic loss and damage experienced by displaced persons (IDMC, 2018; IOM, 2017).

3. Methods

3.1. Contextualizing The Bahamas case in the wider SIDS experience

The Bahamas is an archipelagic Caribbean SIDS consisting of over 700 islands and cays. Hurricanes, tropical storms and depressions are common occurrences in the country with a Bahamian island being affected by a hurricane once every three years (Neely, 2006). The islands are notably flat with elevations close to sea level and are prone to flooding during intense rainfall. Hurricane damages between 2015 and 2017 alone exceeded USD670 million while hurricane impacts between 1980 and 2012 resulted in approximately USD2.5 billion in damages (Smith, 2018b).

The most recent official census for the nation shows a population of just above 350,000 (Government of The Bahamas, 2010a). New Providence, the capital island of the country, is home to approximately 70% of the nation's inhabitants while other residents are spread among 18 other so-called Family Islands. Ragged Island is one of the smaller Family Islands with a population of 72 people (Government of The Bahamas, 2010c) and has been experiencing gradual population decline since 1990 with a net loss of 54 people due to migration of residents, largely to New Providence (Government of The Bahamas, 2010b).

While SIDS vary in terms of particular geographic and socioeconomic characteristics, Ragged Island showcases the attributes that make such small islands across the globe vulnerable to both sudden and slow-onset events associated with climate change. Low levels of development; concentration of the population along the coast; reliance on natural resources for livelihoods such as fishing; presence of particularly vulnerable subgroups including the elderly, children and impoverished communities; small populations; limited economies of scale and low-elevations, are all qualities of Ragged Island that are shared by many SIDS, particularly in rural communities (Hashim & Hashim, 2016; Kelman & West, 2009; Nurse et al., 2014). The frequent experiences of The Bahamas with hurricanes reflects the high exposure of SIDS to tropical storms that has resulted in extensive economic damages in the past, and which are expected to increase significantly as global temperatures rise (Acevedo, 2016). While migration of small island settlements has been part of historic adaptation to environmental change in past centuries, particularly in Pacific islands (Corendea, 2016; Farbotko & Lazrus, 2012), colonization and globalization of SIDS has resulted in a shift of attitudes towards maintaining existing coastal settlements, strong connections to the land and aversion to possibilities of relocation (Nunn & Kumar, 2018). Additionally, as exemplified by Ragged Island, while many SIDS have high rates of urbanization, there are also many rural communities with small populations that are currently experiencing impacts of climate change with insufficient policies or plans to guide responses (Nunn & Kumar, 2018). Therefore, many lessons can be learned from the case study in Ragged Island for SIDS as well as rural, vulnerable communities, particularly how policy gaps and deficiencies can impact Government action and community reaction.

3.2. Assessment of policies, plans, legislation, UNFCCC documents and non-economic loss and damage associated with displacement

National policies, legislation, plans, and UNFCCC documents from The Bahamas that included human mobility, disaster risk reduction or climate change were identified using the authors' familiarity of the policy landscape from years of research in the country. These documents were then reviewed to assess whether displacement or non-economic loss and damage were included and the context in which these issues were addressed. Specifically, each document was reviewed to determine discussion of: (i) human mobility in any form, (ii) displacement in particular, (iii) non-economic loss and damage in any context, and (iv) non-economic loss and damage related to human mobility. The specific details of inclusion of each of these issues as well as the context of their inclusion were also assessed.

Assessment of national policies, legislation and plans is a well-established research methodology that allows for identification of coherence, fragmentation or gaps for specific issues of interest including governance in Caribbean SIDS (Scobie, 2016), approaches to mitigation (Makkonen, Huttunen, Primmer, Repo, & Hildén, 2015), and comparison of how national policies affect emissions rates (Lachapelle & Paterson, 2013). This methodology has also been used to identify countries that include aspects of human mobility and climate change in national UNFCCC documents and has highlighted that there is need for additional research in this area, particularly to cover comprehensive assessments that include disaster risk reduction strategies and internal displacement policies (International Organization for Migration, 2018). News media and reports of the displacement of Ragged Island residents were assessed to determine Government action, community reactions and articulations of non-economic loss and damage that were experienced as a result of this event. The displacement of all residents from Ragged Island was an issue of national concern and so there was extensive news coverage of the plight of residents. Assessment of news media has been proven as a valid methodology for insight into impacts associated with extreme events, with more recent research expanding into utilization of social media as a source of information (Nerlich & Jaspal, 2014; Oh, Kwon, & Raghav Rao, 2010). For small islands in particular, utilizing news media to assess impacts of extreme events has been used in situations where collection of primary data has been infeasible or where assessment of impacts would be delayed by waiting for primary data (Chunara, Andrews, & Brownstein, 2012). As underscored by Tschakert, Ellis, Anderson, Kelly, and Obeng (2019), assessments of non-economic loss and damage is an example of 'slow' research, requiring long-term community engagement, co-production of knowledge and participatory and qualitative methodologies. For this study, we use news media and reports to establish that non-economic loss and damage has taken place and motivated community reaction, and that there is a need for further research on these issues, including through the collection of primary data obtained through interviews or surveys of Ragged Island residents.

4. Results

4.1. Policy, plans, legislation and UNFCCC documents review

Six national policies, plans, legislation or UNFCCC reports on The Bahamas related to human mobility, disaster risk reduction or climate change were identified and reviewed: First National Communication on Climate Change to the UNFCCC (FNC), Second National Communication on Climate Change to the UNFCCC (SNC), National Policy for the Adaptation to Climate Change (NPAC), Disaster Preparedness and Response Act (DPRA), Nationally Determined Contribution (NDC) and National Development Plan (NDP).

The FNC – submitted to the UNFCCC in 2001 – refers to the high vulnerability of the country to climate change impacts and frames the entire report in the context of the close personal relationship between the Bahamian people, the land and the sea, as well as their traditional reliance on natural resources for subsistence (Government of The Bahamas, 2001). The report stresses the traditional livelihoods of Family Island residents including fishing and farming, and that impacts from climate change could cause a 'drastic change in lifestyles' to these communities. Adaptation options include retreat, abandonment or accommodation, although the report notes that relocation may be both 'difficult to visualize' for residents, as well as unfeasible due to the low elevations of land masses. The report notes that, 'it is also clear that in the adaptation process, hard choices will have to be made to abandon some settlements or not replace them when damaged'. No connection between retreat or abandonment and non-economic loss and damage is made in the report except for the above-mentioned reference to 'drastic change in lifestyles' (Government of The Bahamas, 2001, p. 9).

The 2014 SNC does not mention traditional lifestyles except to comment that life in small and isolated Family Islands – such as Ragged Island – has remained 'virtually unchanged over the past century' (Government of The Bahamas, 2014). Although sea, lake and overland surge modelling mentioned in the report does highlight that in worst case scenarios most parts of the islands would be underwater, no mention is made of displacement or any aspects of non-economic loss and damage. Although this SNC was the first national communication of The Bahamas to include a specific vulnerability and adaptation assessment, only the health impacts of climate change, specifically vector-borne diseases, are mentioned in detail. Relocation is listed as an adaptation option for sea level rise, but no discussion of the details or the impacts on communities of this option are provided.

The 2005 NPAC recognizes the high vulnerability of the country to sea level rise and extreme events, as well as the limited human and financial resources available to manage these vulnerabilities (Government of The Bahamas, 2005). Negative health impacts due to the effects of climate change, as well as the exacerbating impacts of extreme events such as hurricanes on existing socio-economic and environmental challenges, are also acknowledged. The NPAC recommends development and implementation of plans for relocation or protection of settlements and infrastructure most at risk to climate change, and also recommends incorporation of climate change into existing or proposed national disaster planning. However, the NPAC does not provide further detail on what the relocation policy could encompass, or what protections it could provide for the preservation of cultural heritage and community values.

The DPRA was passed in 2006, but did not implement the specific directive from the NPAC to incorporate climate change into national disaster planning, and further fails to include any discussion on climate change (Government of The Bahamas, 2008). The Act provides a centralised structure for disaster preparedness and response, establishing the National Emergency Management Agency (NEMA). Each Family Island can appoint a committee with representatives from each settlement that can assist the Director of NEMA with developing disaster preparedness measures and discharging the functions of NEMA. The Director has a broad mandate to implement a general policy focusing on mitigating, preparing for, responding to and recovering from emergencies and disasters and can recommend and develop national policies to promote these broad aims, as well as develop disaster risk assessment maps. The Director can declare a specially vulnerable area and develop a precautionary plan, including strategies, policies and standards for development and maintenance for structures. While the Act focuses on disaster response, it does little to structure disaster preparedness, promote risk reduction or address long-term recovery.

While the more recent NDC submitted to the UNFCCC in 2015 mentions the country's vulnerability to climate change and the desire of the Government to preserve its citizens' way of life, it does not include human mobility or non-economic loss and damage in the context of displacement. Loss and damage are mentioned in the context of loss of energy assets and other infrastructure due to extreme events and loss of potable water resources. Impacts on health due to climate change such as heat-stress and vector-borne diseases are mentioned in the context of including these in national emergency planning and education of health professionals. Additionally, the need to develop a comprehensive national land use and management plan is listed which includes regulation of future settlements with sustainability and potable water requirements in mind, but no details of this project are provided. (Government of The Bahamas, 2015).

The draft NDP assesses gaps and inadequacies of national plans and policies on climate change and disaster risk management (Government of The Bahamas, 2016). It recommends modernization of infrastructure to withstand climate change impacts, development of an integrated national land use plan, and hazard mapping for sustainable planning. The plan also advises revision of the NPAC and passage of laws to implement its policy directives, integration of disaster risk reduction in sustainable development policies and planning, and integration of climate induced migration into national plans and policies. The impact of disasters on communities is mentioned with concern for 'grave consequences for the survival, dignity and livelihood of individuals, particularly the poor' with suggestions to improve early warning and evacuation systems. However, other than this brief comment, human mobility and components of non-economic loss and damage are not included.

4.2. Non-economic loss and damage effects of displacement

At the beginning of September 2017, Hurricane Irma passed over Ragged Island as a category five hurricane leaving the island in ruins with every home and structure destroyed (Russell, 2017). The island was deemed 'unliveable' by the Government, with most infrastructure reduced to rubble. The majority of inhabitants had left the island either before or directly after the storm. At that time, no essential utilities such as power or running water were available, the school and health clinic were in ruins, and the stench of dead animals was overwhelming.

In the week after the passing of Hurricane Irma, the Government made a plane available for removal of the 18 remaining residents. While the evacuation was not mandatory, residents were strongly advised to leave given the dire living conditions and threats to health and wellbeing. Coverage of the evacuation revealed that many residents 'were understandably reluctant to leave the only place they've ever known as home, expressing concern for personal belongings' (Russell, 2017). The Prime Minister, when pressed, would not provide a timeline for reconstruction of the community and restoration of essential services that would affect when displaced residents would be able to return home.

Most of the displaced inhabitants of Ragged Island were rehoused by relatives or friends in New Providence or in the nearby Family Island of Exuma (Russell, 2017). All students were evacuated to New Providence (Lockhart, 2017). None of the inhabitants received government assistance to purchase building materials to rebuild their homes in the months following the storm (Cartwright-Carroll, 2017).

However, refusing to accept the Government assessment that their island home was unliveable, descendants of Ragged Island together with members of the Ragged Island Cultural Heritage Association (RICHA) began collective efforts to start the rebuilding process a few months after the storm (Lockhart, 2017). The islanders, known for their resilience, initiated a 'Repair, Remodel and Restore Initiative' to kick-start the process of returning home by cleaning up debris and removing animal carcasses. While their short-term clean-up work was interrupted by evacuation orders for Hurricane Maria in October 2017, members of RICHA continued their efforts by preparing care packages for evacuated residents and spearheading efforts to raise money for reconstruction and restoration efforts.

Despite community efforts, months after Hurricane Irma ravaged Ragged Island, the island was 'still in ruins' (Cartwright-Carroll, 2017). In December 2017, an official from the Inter-American Development Bank (IDB) stated that 'it doesn't make sense' to rebuild on Ragged Island due to exorbitant costs (Smith, 2017). This suggestion was dismissed by the Minister of Finance as well as by the Parliamentary representative for Ragged Island who stated, ' … the inalienable rights of the people of Ragged Island to live in their ancestral home are not something The Bahamas should entertain any compromise on' (Cartwright-Carroll, 2017). However, eight months after the hurricane passed, no plans had been implemented to restore the island (Brown, 2018). While the Government had publicly committed to making Ragged Island the first green island in the region, despite its public rhetoric stating its commitment to rebuild, by May 2018 the school, clinic and police station had still not been rebuilt.

Faced with Government inaction on the ground, Ragged Islanders therefore continued to resort to their own community and descendants for restoration and reconstruction efforts. Approximately 60 inhabitants had returned to the island by May 2018, despite no critical services being restored to the island and the Government not lifting the 'uninhabitable' declaration (Adderley, 2018a). During this time members of the Restoration Ragged Island Association (RRIA) called on the Government to rebuild critical infrastructure so the islanders could return to a semblance of normality. The President of RRIA stated, 'Ragged Islanders are a proud people. We don't ask for handouts; we ask for hand-ups' (Smith, 2018a). Almost a year after the storm in August 2018, the Government made financial contributions to inhabitants for damage to their homes. Although financial compensation was given, pointed comments were made by the Director of NEMA who noted that almost half of the homes destroyed by the storm were not built in accordance with the national Building Code, emphasizing the importance of adherence to the Building Code (NEMA, 2018).

By August 2018, residents continued to be frustrated by the lack of Government action in restoration of critical services, which they claimed was inhibiting local efforts to rebuild and restore their community (Adderley, 2018a). The RRIA purchased a reverse osmosis plant to provide more consistent potable water to the islanders. Acceding to their requests, in spite of the large number of infrastructural facilities that needed to be repaired after the 2017 hurricane season, in September 2018 the Government approved restoration work to key government infrastructure such as the local administrator's office, the health clinic and police station (Adderley, 2018b). When this work will be completed is unclear, although it means that in principle, life for the remaining community members should be restored to normality in the future, and they will be able to regain at least part of their community life once again. The importance of extended community members in the struggle to gain the attention and scarce resources of the Government is highlighted here, along with the determination of a small community to restore their way of life. In the face of Government inaction and delayed compensation, the community successfully challenged the formal assessment by the Government that their island was uninhabitable, and the opinion of a regional bank that it was not financially prudent to rebuild a community. The lack of assistance by the Government to the community also highlights the inconsistency between the Government's public rhetoric of support for rebuilding the community and its actions, perhaps demonstrating the implications of the absence of a coherent, systematic policy response on the part of the Government.

5. Discussion

5.1. National policy landscape

The national policy landscape in the context of climate-induced displacement and non-economic loss and damage in The Bahamas is scarce, outdated and fragmented. Only the FNC refers to climate-induced displacement in any detail, and only the NPAC recommends the development of a relocation policy. None of the policies plans or legislation refer to components of non-economic loss and damage, except for references to health impacts due to climate change – specifically vector-borne diseases – and loss of traditional lifestyles. Additionally, none of the documents connect displacement with non-economic loss and damage. While earlier reports acknowledged and explored the vulnerability of traditional lifestyles, and the complexities of relocation as an adaptation option, they did not make the connection between these issues. Although some national plans, policies and legislation acknowledge potentially severe consequences of extreme events and climate-induced displacement, no policy approach to non-economic loss and damage was developed. Policy documents instead anticipated the need for new or revised policies and therefore acknowledged existing policy gaps and deficiencies in the context of human mobility and non-economic loss and damage.

In terms of a policy that specifically addresses human mobility, the NPAC does recommend formalisation of a relocation policy, and the DPRA provides a broad mandate to NEMA to develop such a policy. However, these recommendations have yet to be taken up by the Government and no formal policy has been developed or implemented. There appears to be some progress on this front as the Department of Housing has an informal policy of not rebuilding in vulnerable areas and relocating impacted communities to higher ground (Lacambra et al., 2018). However, this policy has not yet been approved and is not available to the public. Recognizing the gaps in current policies, recommendations have been made for the country to develop ex-ante post-disaster recovery plans with the explicit objectives of reducing pre-existing levels of vulnerability, including in the re-establishment of livelihoods and potential relocation of communities (Lacambra et al., 2018).

When considering the case of The Bahamas on a global scale, it is not surprising that this SIDS has not developed a robust policy response to climate-induced displacement and non-economic loss and damage, despite experiencing challenges in the present term. On a global scale, while over 140 countries and territories were affected by displacement in 2017, only 31 currently have dedicated policies or strategies focused on internal displacement (Internal Displacement Monitoring Centre, 2018). In a review of connections between human mobility and climate change at the national level, the International Organization for Migration (2018) assesses how these issues are included in national policies, strategies and legal frameworks. Their findings highlight lack of policy coherence and coordination at the national scale: some climate change policies include human mobility while corresponding human mobility policies do not include climate change; human mobility is included in some but not all national climate change policies (e.g. in national communications but not in NDCs); and many countries lack specific policies, plans or legislation that address human mobility.

For SIDS in particular, while those in the Pacific have made some strides in addressing climate-induced migration and displacement through policy development, Caribbean SIDS have comparatively not made much progress (Thomas & Benjamin, 2018). The capacity constraints of SIDS to develop and implement effective policies to address climate-induced displacement and loss and damage include lack of collection and analysis of nationally scaled data; limited studies on SIDS-specific human mobility; understaffed national institutions and lack of funding (Thomas & Benjamin, 2017, 2018). Additionally, Benjamin et al. (2018) suggest that SIDS are trapped in an 'unvirtuous cycle' where costly climate impacts worsen capacity constraints and require countries to spend limited funds on responding to impacts rather than allocating resources to building resilience-such as through the development and implementation of robust policies.

However, in the absence of policies or plans to guide displacement, movement of residents due to flooding and hurricane damage is often done in an ad-hoc manner with little regard paid to impacts of these moves on residents or to the preferences or desires of communities (Wilkinson, 2018). The Ragged Island example illustrates the negative impacts of acting in such a policy void. The Government adopted an ad-hoc relocation approach and the community reacted by expressing, and acting on, a strong desire to return home. While the Government respected the desires of the community, they did not facilitate their return in the face of the high economic costs of rebuilding and a reluctance expressed by regional funding bodies to finance reconstruction efforts. In the face of a policy void on the issue, Government action was delayed and inadequate post-evacuation, and it appears that little or no consultation with communities took place. In contrast, community reaction to Government inaction was swift, consistent and deliberately focused on rebuilding and therefore preserving their way of life and sense of place.

5.2. Non-economic loss and damage

Non-economic loss and damage stemming from the prolonged displacement of residents of Ragged Island is evident, in particular on health. Potential implications on health of the residents was a major impetus for the advised evacuation of the island in the week following the hurricane. The lack of running water and electricity, destruction of the health clinic, absence of medical personnel, presence of animal carcasses, and threats to water supply resulted in abominable living and health conditions. However, while health risks brought about by continued occupation of the island were acknowledged by the Government (Russell, 2017), there have been no follow-up studies on residents' health after they left the island.

Previous research has shown that displaced persons residing in temporary homes often face additional health threats (Chen et al., 2016). Displacement and climate-induced impacts can lead to stress, loss, grief, hopelessness and alienation, leading to negative impacts on health and well-being (Tschakert et al., 2017). These threats to the health of Ragged Islanders are further compounded by the return of some residents in the weeks following the storm to begin the recovery and restoration process, despite the lack of running water and electricity. With 60 people living on the island by May 2018 -despite the continued lack of utilities and the island still being deemed uninhabitable – there may be health effects on residents that are going unrecorded.

The continued efforts of Ragged Islanders to return to their homes shows their strong sense of place and connection to the island that was disrupted by their prolonged displacement. After waiting for months for action by the Government, residents along with a larger group of supporters spearheaded the recovery and reconstruction process and began to move back to the island, despite lack of Governmental services. This strong connection along with the rights of the residents to return to their homes was acknowledged by the Government. However, the non-economic consequences of their extended displacement due to lack of Government action has not been acknowledged. While residents were eventually provided with some financial support to rebuild their homes, the effects of the prolonged displacement on residents' sense of place and connection to the island and community have not been addressed.

The displacement highlighted the strong social network of Ragged Islanders; their sense of agency; and their strong sense of connection to their home and to their community members. The tenacity of Ragged Islanders resulted in return to their island despite many obstacles and lack of Government support. However, without these strong characteristics, it is clear that the displacement would have lasted even longer. As the Government only approved restoration work in September 2018 with no indication of when the work will be complete, without the concerted effort of Ragged Islanders along with their broader social networks, the island would have likely remained uninhabited for a more prolonged period of time. Such an extended displacement would result in further non-economic loss and damage including potential loss of culture and disconnection of the community.

In this instance, the community determined that their way of life and sense of place, social cohesion and identity was worth preserving in the face of extreme impacts and continued risks of climate change. Such a place-attachment can be a motivator and predictor for community engagement with climate change adaptation efforts (Tschakert et al., 2017), and how the community will engage and make decisions in the face of climate-induced loss and damage. Ignoring this sense of place led to an inadequate Government response. A values-based approach to adaptation which links to understandings of place, well-being and lived experience can be an effective tool and offers guidance for development of policies (Tschakert et al., 2017).

5.3. Implications for rebuilding

The lack of acknowledgement and assessment of non-economic loss and damage experienced by Ragged Islanders means that these issues were not taken into consideration when evaluating the costs of rebuilding. Failure to include consideration of health impacts, potential loss of culture, loss of sense of place and connections to the island and community result in a skewed assessment of the costs and benefits associated with continued occupation of the island. Communities that may have limited values of economic assets and small populations face assessments that rebuilding of their homes is infeasible, as was the case for Ragged Island (Smith, 2017).

While the country has a Building Code, it is often not enforced, and lack of adherence to the Building Code may have delayed Government reimbursement of building costs as illustrated by the comments made by NEMA. It is unclear whether new structures rebuilt by the community are compliant with the Building Code, and it appears that no or limited consultation between communities and the Government has been carried out or their potential continued vulnerability to the impacts of climate change due to their rebuilding efforts. The community relied on its own resources to rebuild as a mechanism to preserve Ragged Islanders' way of life and sense of place, which gave meaning and value to their lives, and a sense of well-being which exceeded, for them, the costs of rebuilding and exposure to further risks.

The reliance of SIDS on international aid and funding to rebuild means that skewed perceptions of the costs and benefits of rebuilding by international funders may result in small island communities being incapable of returning to their homes after disaster strikes. Without policies that include non-economic loss and damage, Governments and international agencies develop their own recovery plans which focus on physical reconstruction without inclusion of issues such as dignity and the desire of residents to feel 'at home' (Wilkinson, 2018). The approach of the Inter-American Development Bank in the case of Ragged Island highlights this exclusive focus on economic costs, and minimizes the value and importance of non-economic loss and damage to affected communities. The nature of the policy environment thus has significant implications on support for rebuilding. Olawuyi (2016) notes that climate change mitigation and adaptation measures can result in land grabs, forced displacement, marginalisation, exclusion and governmental repression in developing countries. Failure to include assessments of non-economic loss and damage would lend further support to these infractions of justice. The Ragged Island case study illustrates the need for Governments to consult with communities post-disaster on rebuilding and relocation efforts and to include non-economic loss and damage in evaluating the costs of rebuilding.

6. Conclusion

The prolonged displacement of Ragged Islanders, and the economic and non-economic loss and damage that the residents experienced, highlights the need for robust and effective policies, plans or strategies that address these issues. As this case shows, without clear guidance and acknowledgement of the many effects of displacement for communities, Government action in assisting residents to either return to their homes or to relocate to less vulnerable areas is delayed, resulting in extended lengths of displacement for residents who are anxious to return to some sense of normality, and extending the non-economic loss and damage that is incurred.

This study of The Bahamas provides useful lessons for other Caribbean SIDS and for SIDS in other regions. As impacts of climate change increase, SIDS are at the forefront of experiencing loss and damage in all of its forms (Thomas, Schleussner, & Kumar, 2018). In particular, the frequency of intense tropical cyclones is expected to increase as global temperatures rise, leading to heightened loss and damage associated with these extreme events in SIDS. This study illustrates how important a sense of place and a desire to rebuild and return is to vulnerable communities, even if these efforts place communities in a more vulnerable situation. The community itself delineated and assessed limits to Government adaptation efforts using the lens of non-economic loss and damage. The community asserted the importance of nuanced, place-specific and culturally relevant losses, which they valued over and above physical and economic losses. Government policies should do the same by considering and incorporating non-economic loss and damage in national policies and consulting with affected communities. Culturally specific values could be a driver of vulnerability and may reverse or counter adaptation efforts (Tschakert et al., 2017).

While SIDS have articulated that guidance from the UNFCCC is needed to advance national policies on loss and damage, it is clear that the absence of national policies has significant implications for sustainable development and well-being of communities in the present term. For example, Barbuda was also completely evacuated in 2017 after Hurricane Maria, resulting in the island being uninhabited for the first time in 300 years. Antigua and Barbuda have also struggled with a timely return of residents to the island and the displacement has exposed a number of political, financial, environmental and social challenges that are impeding action (Serdeczny, Menke, et al., 2018).

Outcomes from both COP 23 and COP 24 in 2018 underscore the importance of SIDS in taking the lead on developing national policies, plans and strategies for loss and damage, rather than awaiting guidance from the UNFCCC. At COP23, agreed outcomes fell far short of the key issues that SIDS advocated for, with no reference to finance for incurred loss and damage, failure to include loss and damage as a permanent agenda item in negotiations and no provisions for adequate financing for the work of the WIM ExCom (Benjamin et al., 2018). At COP 24, loss and damage was reflected in the transparency framework which means that countries will be able to officially report current and projected impacts, activities to address loss and damage, institutional arrangements to implement such activities, and support needed (Serdeczny, Pierre-Nathoniel, & Siegele, 2018). Thus, countries will need to ensure that they have the national capacities to identify, assess and report on loss and damage – in all of its forms – to facilitate access to support. This case study illustrates the urgent need for SIDS to develop their own national policies on non-economic loss and damage.

Disclosure statement

No potential conflict of interest was reported by the authors.

Funding

This work was supported by the German Federal Ministry for the Environment, Nature Conservation, and Nuclear Safety and the Killam Trust.

ORCID

Adelle Thomas ⓘ http://orcid.org/0000-0002-0407-2891
Lisa Benjamin ⓘ http://orcid.org/0000-0001-9696-9817

References

Acevedo, S. (2016). *Gone with the wind: Estimating Hurricane and climate change costs in the Caribbean authorized for distribution.* Retrieved from https://www.imf.org/external/pubs/ft/wp/2016/wp16199.pdf
Adderley, M. (2018a). NEMA – Ragged Island still uninhabitable. *The Tribune*, p. 1. Retrieved from http://www.tribune242.com/news/2018/may/30/nema-ragged-island-still-uninhabitable/
Adderley, M. (2018b). Vital ragged Island facilities to be rebuilt. *The Tribune*, p. 1. Retrieved from http://www.tribune242.com/news/2018/sep/07/vital-ragged-island-facilities-be-rebuilt/
Benjamin, L., Thomas, A., & Haynes, R. (2018). An 'Islands' COP'? Loss and damage at Cop23. *Review of European, Comparative and International Environmental Law, 27,* 332–340.
Brown, S. (2018). Govt. still committed to ragged Island. *EyeWitness News*, p. 1. Retrieved from https://ewnews.com/govt-still-committed-to-ragged-island
Cartwright-Carroll, T. (2017). Ragged Island still in ruins. *The Nassau Guardian*, p. 1. Retrieved from https://thenassauguardian.com/2017/12/06/ragged-island-still-ruins/
Cashman, A., & Nagdee, M. R. (2017). Impacts of climate change on settlements and infrastructure in the coastal and Marine environments of Caribbean small island developing states (SIDS). *Science Review, 2017,* 155–173. Retrieved from http://crfm.int/
Chen, B., Halliday, T. J., & Fan, V. Y. (2016). The impact of internal displacement on child mortality in post-earthquake Haiti: A difference-in-differences analysis. Retrieved from https://equityhealthj.biomedcentral.com/track/pdf/10.1186/s12939-016-0403-z
Chunara, R., Andrews, J. R., & Brownstein, J. S. (2012). Social and news media enable estimation of epidemiological patterns early in the 2010 Haitian Cholera outbreak. *The American Journal of Tropical Medicine and Hygiene, 86*(1), 39–45. Retrieved from http://www.ncbi.nlm.nih.gov/pubmed/22232449
Corendea, C. (2016). *Climate law and governance working paper series: Development implications of climate change and migration in the Pacific.* Montreal. Retrieved from http://www.cisdl.org
CRED. (2019). Emergency Events Database (EM-DAT) | Centre for Research on the Epidemiology of Disasters. Retrieved from https://www.cred.be/projects/EM-DAT
Curtain, R., & Dornan, M. (2019). *A pressure release valve? Migration and climate change in Kiribati, Nauru and Tuvalu.* Australia. Retrieved from http://devpolicy.org/publications/reports/Migration-climate change-Kiribati-Nauru-Tuvalu.pdf
ECLAC. (2018). *Irma and Maria by numbers.* Port of Spain. Retrieved from www.eclac.org/portofspain
Farbotko, C., & Lazrus, H. (2012). The first climate refugees? Contesting global narratives of climate change in Tuvalu. *Global Environmental Change, 22*(2), 382–390.
Government of Dominica. (2018). *2018 budget: From survival to sustainability and success: A resilient Dominica.* Roseau.
Government of The Bahamas. (2001). *First national communication on climate change.* Nassau. Retrieved from https://unfccc.int/resource/docs/natc/bahnc1.pdf
Government of The Bahamas. (2005). Science and technology commission. *National Policy for the Adaptation to Climate Change.* Nassau. Retrieved from www.grida.no/climate/ipcc/aviation/in-dex.htm
Government of The Bahamas. (2008). *Disaster preparedness and response Act.* Nassau. Retrieved from http://laws.bahamas.gov.bs/cms/images/LEGISLATION/PRINCIPAL/2006/2006-0004/DisasterPreparednessandResponseAct_1.pdf
Government of The Bahamas. (2010a). *Census of population and housing.* Nassau. Retrieved from https://www.bahamas.gov.bs/wps/wcm/connect/a6761484-9fa0-421d-a745-34c706049a88/Microsoft+Word+-+2010+CENSUS+FIRST+RELEASE+REPORT.pdf?MOD=AJPERES
Government of The Bahamas. (2010b). *Internal migration.* Nassau. Retrieved from https://www.bahamas.gov.bs/wps/wcm/connect/d5772283-6671-40e3-96b9-785cf92ae2bb/2010+Internal+Migration.pdf?MOD=AJPERES
Government of The Bahamas. (2010c). *Ragged Island population by settlement.* Nassau. Retrieved from https://www.bahamas.gov.bs/wps/wcm/connect/0f9f1ede-5069-4c16-83dd-a0c9a3a9e701/RAGGED+ISLAND+POPULATION+BY+SETTLEMENT_2010+CENSUS.pdf?MOD=AJPERES

Government of The Bahamas. (2014). *Second national communication report of the commonwealth of The Bahamas*. Nassau. Retrieved from https://unfccc.int/sites/default/files/resource/bhsnc2.pdf

Government of The Bahamas. (2015). *Intended nationally determined contribution*. Nassau. Retrieved from https://www4.unfccc.int/sites/ndcstaging/PublishedDocuments/Bahamas First/Bahamas_COP-22 UNFCCC.pdf

Government of The Bahamas. (2016). *The national development plan of The Bahamas*. Nassau. Retrieved from http://www.vision2040bahamas.org/media/uploads/Draft__National_Development_Plan_01.12.2016_for_public_release.pdf

Hamza, M., Koch, I., & Plewa, M. (2017). Disaster-induced displacement in the Caribbean and the Pacific. *Forced Migration Review, 56*, 62–64. Retrieved from https://www.imf.org/external/np/pp/eng/2016/110416.pdf

Hashim, J. H., & Hashim, Z. (2016). Climate change, extreme Weather events, and human health implications in the Asia Pacific region. *Asia Pacific Journal of Public Health, 28*(2_suppl), 8S–14S. Retrieved from http://www.ncbi.nlm.nih.gov/pubmed/26377857

Hill, M. C. (2016). Closing the gap: Towards rights-based protection for climate displacement in Low-lying small Island States. *New Zealand Journal of Environmental Law, 20*, 43–75. Retrieved from https://heinonline.org/HOL/Page?handle=hein

IDMC. (2018). *Global report on internal displacement 2018*. Geneva. Retrieved from http://www.internal-displacement.org/global-report/grid2018/downloads/report/2018-GRID-spotlight-atlantic-hurricane-season.pdf

Internal Displacement Monitoring Centre. (2018). *Global report on internal displacement*. Retrieved from http://www.internal-displacement.org/sites/default/files/publications/documents/201805-final-GRID-2018_0.pdf

International Organization for Migration. (2018). *Mapping human mobility and climate change in relevant national policies and institutional frameworks international organization for migration (IOM) Task Force on Displacement Activity I.1 The Warsaw international mechanism for loss and damage associated W*. Retrieved from https://unfccc.int/sites/default/files/resource/WIM TFD I.1 Output.pdf

IOM. (2017). *Displacement tracking matrix (DTM)-Antigua*. Geneva. Retrieved from https://reliefweb.int/sites/reliefweb.int/files/resources/Return Intention Survey Report_V1_ Antigua Barbuda %2800%29.pdf

Kaenzig, R., & Piguet, E. (2014). Migration and climate change in Latin America and the Caribbean. In E. Piguet & F. Laczko (Eds.), *People on the move in a changing climate : The regional impact of environmental change on migration* (Vol. 2, pp. 155–176). Dordrecht: Springer Netherlands. Retrieved from http://www.springerlink.com

Kelman, I., & West, J. J. (2009). Climate change and small island developing States: A critical review. *Ecological and Environmental Anthropology, 5*, 1. Retrieved from http://www.ilankelman.org/contact.html

Lacambra, S., et al. (2018). *Index of governance and public policy in disaster risk management (IGOPP) National Report for The Bahamas*. Retrieved from http://www.iadb.org

Lachapelle, E., & Paterson, M. (2013). Drivers of national climate policy. *Climate Policy, 13*(5), 547–571. Retrieved from http://www.tandfonline.com/doi/abs/10.1080/14693062.2013.811333

Lewsey, C., Cid, G., & Kruse, E. (2004). Assessing climate change impacts on coastal infrastructure in the Eastern Caribbean. *Marine Policy, 28*(5), 393–409. Retrieved from https://www.sciencedirect.com/science/article/pii/S0308597X03001313

Lockhart, S. (2017). Ragged Islanders Determined in their rebuilding commitment, Says administrator. *The Freeport News*, p. 1. Retrieved from http://thefreeportnews.com/news/local/ragged-islanders-determined-rebuilding-commitment-says-administrator/

Makkonen, M., Huttunen, S., Primmer, E., Repo, A., & Hildén, M. (2015). Policy coherence in climate change mitigation: An ecosystem Service approach to Forests as Carbon Sinks and Bioenergy Sources. *Forest Policy and Economics, 50*, 153–162. Retrieved from http://dx.doi.org/10.1016/j.forpol.2014.09.003

Méheux, K., Bank, W., Dominey-Howes, D., & Lloyd, K. (2007). Natural hazard impacts in small island developing states: A review of current knowledge and future research Needs. *Natural Hazards, 40*, 429–446. Retrieved from https://www.researchgate.net/publication/227296545

Millar, I. (2007). There's no place like home: Human displacement and climate change. *Australian International Law Journal, 14*, 71. Retrieved from https://heinonline.org/HOL/Page?handle=hein

Monioudi, I. N., Asariotis, R., Becker, A., Bhat, C., Dowding-Gooden, D., Esteban, M., … Witkop, R. (2018). Climate change impacts on critical international transportation assets of Caribbean small island developing states (SIDS): The case of Jamaica and Saint Lucia. *Regional Environmental Change, 18*(8), 2211–2225. Retrieved from http://link.springer.com/10.1007/s10113-018-1360-4

Mortreux, C., & Barnett, J. (2009). Climate change, migration and adaptation in Funafuti, Tuvalu. *Global Environmental Change, 19*(1), 105–112. Retrieved from http://linkinghub.elsevier.com/retrieve/pii/S0959378008000903

Neely, W. (2006). *The major hurricanes to affect the Bahamas : Personal recollections of some of the greatest storms to affect the Bahamas*. Bloomington: AuthorHouse. Retrieved from https://www.barnesandnoble.com/w/the-major-hurricanes-to-affect-the-bahamas-wayne-neely/1119661531

NEMA. (2018). NEMA Makes financial contributions to hurricane Irma victims on Ragged island. *Bahamas Information Services, 1*, 1–1. Retrieved from http://www.bahamas.gov.bs/wps/portal/public/gov/government/news/nema makes financial contributions to hurricane irma victims on ragged island.

Nerlich, B., & Jaspal, R. (2014). Images of extreme weather: Symbolising human responses to climate change. *Science as Culture, 23*(2), 253–276. Retrieved from http://www.tandfonline.com/doi/abs/10.1080/09505431.2013.846311

Nunn, P., & Kumar, R. (2018). Understanding climate-human Interactions in small island developing states (SIDS): Implications for future livelihood sustainability. *International Journal of Climate Change Strategies and Management, 10*(2), 245–271. Retrieved from https://doi.org/10.1108/IJCCSM-01-2017-0012

Nurse, L. A. (2014). Small islands. In K. L. Ebi Barros, V. R. C, B. Field, D. J. Dokken, M. D. Mastrandrea, K. J. Mach, & T. E. Bilir (Eds.), *Climate change 2014: Impacts, adaptation, and vulnerability. Part B: Regional aspects. Contribution of working group II to the fifth assessment report of the intergovernmental panel on climate change* (Vol. 42, pp. 1613–1654). Cambridge: Cambridge University Press. Retrieved from https://www.ipcc.ch/site/assets/uploads/2018/02/WGIIAR5-Chap29_FINAL.pdf

Oh, O., Kwon, K. H., & Raghav Rao, H. (2010). An exploration of social media in extreme events : Rumors theory and Twitter during the Haiti earthquake 2010. *Thirsty First International Conference on Information Systems, St. Louis*: 231. Retrieved from https://www.researchgate.net/publication/221599216

Olawuyi, D. (2016). Advancing climate justice in national climate actions: The promise and limitations of the United Nations human rights-based approach. In R. S. Abate & D. C. Washington (Eds.), *CLIMATE JUSTICE case studies in global and regional governance challenges* (pp. 3–24). Washington, DC: Environmental Law Institute. Retrieved from www.ajmilesmedia.com

Russell, K. (2017). Unliveable: PM urges remaining ragged Island residents to evacuate. *The Tribune*, p. 1. Retrieved from http://www.tribune242.com/news/2017/sep/12/unliveable-pm-urges-remaining-ragged-island-reside/

Scobie, M. (2016). Policy coherence in climate governance in Caribbean small island developing states. *Environmental Science & Policy*, *58*, 16–28. Retrieved from https://www.sciencedirect.com/science/article/pii/S1462901115301258

Serdeczny, O., Menke, I., & Thomas, A. (2018). How to ensure solutions really work – key questions for the Suva expert dialogue on loss and damage. *Climate Analytics*, p. 1. Retrieved from https://climateanalytics.org/blog/2018/how-to-ensure-solutions-really-work-key-questions-for-the-suva-expert-dialogue-on-loss-and-damage/

Serdeczny, O., Pierre-Nathoniel, D., & Siegele, L. (2018). Progress on loss and damage in Katowice. *Climate Analytics*, p. 1. Retrieved from https://climateanalytics.org/blog/2018/progress-on-loss-and-damage-in-katowice/

Serdeczny, O., Waters, E., & Chan, S. (2016). *Non-economic loss and damage: Addressing the forgotten side of climate change impacts*. Bonn. Retrieved from https://www.die-gdi.de/uploads/media/BP_3.2016_neu.pdf

Smith, S. (2018a). Ragged Island Assoc. Urges Govt to Act – The Nassau Guardian. *The Nassau Guardian*, p. 1. Retrieved from https://thenassauguardian.com/2018/05/01/ragged-island-assoc-urges-govt-to-act/

Smith, X. (2017). 'It doesn't make sense' to rebuild Ragged Island, Says IDB Official. *The Nassau Guardian*, p. 1. Retrieved from https://thenassauguardian.com/2017/12/05/doesnt-make-sense-rebuild-ragged-island-says-idb-official/

Smith, X. (2018b). "Palacious: Between 2015 and 2017 Hurricanes Cost $678 Mil. In Losses. *The Nassau Guardian*, p. 1. Retrieved from https://thenassauguardian.com/2018/01/19/palacious-between-2015-and-2017-hurricanes-cost-678-mil-in-losses/

Sullivan, M. (2017). *CRS INSIGHT Hurricanes Irma and Maria: Impact on Caribbean Countries and Foreign Territories*. Retrieved from https://www.usaid.gov/irma.

Thomas, A., & Benjamin, L. (2017). Management of loss and damage in small island developing states: Implications for a 1.5°C or Warmer world. *Regional Environmental Change*, 1–10. doi:10.1007/s10113-017-1184-7

Thomas, A., & Benjamin, L. (2018). Policies and mechanisms to address climate-induced migration and displacement in Pacific and Caribbean small island developing states. *International Journal of Climate Change Strategies and Management*, *10*, 1.

Thomas, A., Schleussner, C.-F., & Kumar, M. (2018). Small island developing states and 1.5 °C. *Regional Environmental Change*, *18*, 8.

Tschakert, P., Barnett, J., Ellis, N., Lawrence, C., Tuana, N., New, M., … Pannell, D. (2017). Climate change and loss, as if people mattered: Values, places, and experiences. *Wiley Interdisciplinary Reviews: Climate Change*, *8*(5), 1–19.

Tschakert, P., Ellis, N. R., Anderson, C., Kelly, A., & Obeng, J. (2019). One thousand ways to experience loss: A systematic analysis of climate-related Intangible Harm from around the world. *Global Environmental Change*, *55*(July 2018), 58–72. Retrieved from https://linkinghub.elsevier.com/retrieve/pii/S0959378018308276

UNFCCC. (2013). *Non-economic losses in the context of the work programme on loss and damage/2 2*. Retrieved from https://unfccc.int/resource/docs/2013/tp/02.pdf

United Nations Office for the Coordination of Humanitarian Affairs. (2017). *OCHA on message : Internal displacement*. New York.

Wilkinson, E. (2018). Towards a more resilient Caribbean after the 2017 Hurricanes. *Overseas Development Institute*: 6. Retrieved from https://sustainabledevelopment.un.org/content/documents/2978BEconcept.pdf

Loss and damage in the IPCC Fifth Assessment Report (Working Group II): a text-mining analysis

Kees van der Geest and Koko Warner

ABSTRACT

'Losses and damages' refer to impacts of climate change that have not been, or cannot be, avoided through mitigation and adaptation efforts. After the establishment of the Warsaw International Mechanism for Loss and Damage (WIM), Loss and Damage is now considered the third pillar – besides mitigation and adaptation – of climate action under the United Nations Framework Convention on Climate Change (UNFCCC). This paper studies what the Contribution of Working Group II to the Fifth Assessment Report of the Intergovernmental Panel on Climate Change (IPCC WGII AR5) has to say about this emerging topic. We use qualitative data analysis software (text mining) to assess which climatic stressors, impact sectors and regions the report primarily associates with losses and damages, and compare this with the focus areas of the WIM. The study reveals that IPCC WGII AR5 primarily associates losses and damages with extreme weather events and economic impacts, and treats it primarily as a future risk. Present-day losses and damages from slow-onset processes and non-economic losses receive much less attention. Also, surprisingly, AR5 has more to say about losses and damages in high-income regions than in regions that are most at risk, such as small island states and least developed countries. The paper concludes with recommendations to the IPCC for its 6th Assessment Report (AR6) to include more evidence on losses and damages from slow-onset processes, non-economic losses and damages and losses and damages in vulnerable countries.

Key policy insights
- IPCC WGII AR5 discusses evidence about losses and damages predominantly in relation to sudden-onset disasters and economic costs.
- More research is needed on losses and damages from slow-onset processes and non-economic loss and damage, particularly in vulnerable countries in the Global South.
- Funding agencies should support research in these areas and IPCC WGII AR6 should pay more attention to these topics.
- Losses and damages are not only a future risk, but already a present-day reality for vulnerable people in climate hotspots. People-centred research by social scientists is crucial for enhancing understanding of what losses and damages mean in the real world.

Introduction

Loss and Damage is an emerging concept in the climate change negotiations, as well as in research, policy and implementation of climate change action, and is expected to grow in importance in the coming years (Mechler et al., 2019). Losses and damages refer to impacts of climate-related stressors that have not been, or cannot be, avoided through mitigation and adaptation efforts (Warner & van der Geest, 2013). Enhanced efforts to cut

---

This article has been republished with minor changes. These changes do not impact the academic content of the article.

greenhouse gas (GHG) emissions, and effective adaptation and risk reduction measures, can reduce future losses and damages, but some losses and damages are unavoidable (Huq, Roberts, & Fenton, 2013; Roberts, van der Geest, Warner, & Andrei, 2014; van der Geest & Warner, 2015). Some studies distinguish explicitly between losses – impacts that are permanent – and damages – impacts that are reversible (Doelle & Seck, 2019; McNamara & Jackson, 2019; Tschakert, Ellis, Anderson, Kelly, & Obeng, 2019). However, in the climate negotiations and in the emerging literature on losses and damages, the term is usually is treated as one single concept (Fankhauser, Dietz, & Gradwell, 2014). Following Byrnes and Surminski (2019), we use the plural form and lower case letters – losses and damages – to refer to impacts beyond or despite adaptation, and the upper case singular form – Loss and Damage – to refer to the associated policy debate.

Efforts to reduce GHG emissions have been insufficient so far, putting the world on a trajectory towards a strong increase in global temperature and associated changes in weather patterns, including precipitation and heat extremes, with high risks for human development (Meinshausen et al., 2009; van Vliet et al., 2012; Schellnhuber et al., 2012). Support for adaptation and risk reduction, particularly in developing countries that are most vulnerable to climate change impacts, has increased over the past decade, but vast adaptation deficits still exist (Burton, 2009). There is a growing consensus that there are constraints and limits to adaptation and the ability to avoid losses and damages (Dow et al., 2013; Preston, Dow, & Berkhout, 2013; Warner, van der Geest, & Kreft, 2013). This was recognized in the Summary for Policy Makers (SPM) of the contribution of Working Group II to the Fifth Assessment Report of the Intergovernmental Panel on Climate Change (IPCC) (henceforth referred to as WGII AR5), which states: 'Under all assessed scenarios for adaptation and mitigation, some risk from adverse impacts remain (very high confidence)' (IPCC, 2014, Summary for Policy Makers, p. 14).

WGII AR5 has, for the first time in an IPCC report, a chapter on adaptation opportunities, limits and constraints (Chapter 16). This chapter is an important input to the Loss and Damage debate, as it focuses on situations in which mitigation and adaptation efforts are not enough to avoid impacts from climate change (Nalau & Leal Filho, 2018). The chapter was added after it was realized in the Fourth Assessment Report (IPCC, 2007) that this had become a reality. The chapter documents existing evidence on factors that make it harder to plan and implement adaptation (constraints) and the points at which actors' objectives cannot be protected from intolerable risks through adaptive actions (limits). When actors face 'hard limits', such adaptive actions are simply not possible. In the case of soft limits, options are *currently* not available (IPCC, 2014).

When actors experience constraints to adaptation, future losses and damages can be avoided, or at least reduced, by addressing these constraints. By contrast, when actors face hard adaptation limits, losses and damages are unavoidable. Besides avoidable and unavoidable, there is a third category, namely unavoided losses and damages (Verheyen & Roderick, 2008). This last category moves the concept from an unsecure future to the present-day realities of vulnerable people. While questions remain about the degree to which losses and damages from extreme weather events can be attributed to global warming (Bouwer, 2011; Huggel, Stone, Auffhammer, & Hansen, 2013; Hulme, 2014; James et al., 2014), it is increasingly clear that climate-related stressors have the potential to cause havoc among populations whose underlying vulnerabilities are not sufficiently addressed by adaptation and risk reduction policy (Roberts & Pelling, 2019).

The concept of Loss and Damage first emerged in the climate negotiations in the early 1990s, when the Alliance of Small Island States (AOSIS) called for an insurance pool to compensate low-lying developing countries for the losses and damages caused by sea level rise. After this, it took more than two decades before the concept was institutionalized under the UNFCCC (Calliari, Surminski, & Mysiak, 2019). This happened at the 19th Conference of the Parties (COP 19) in 2013 with the establishment of the Warsaw International Mechanism for Loss and Damage associated with Climate Change Impacts (WIM). UNFCCC decision 2/CP.19 to establish the WIM acknowledges that losses and damages can be reduced by adaptation and risk management strategies. However, it also recognizes that losses and damages sometimes involve more than what can be adapted to or in other words, that some losses and damages cannot be avoided (UNFCCC, 2013a).

The objective of the WIM is to address Loss and Damage associated with impacts of climate change, including extreme events and slow onset events in developing countries that are particularly vulnerable to the adverse effects of climate change. It has three functions: (i) Enhancing knowledge and understanding; (ii) strengthening dialogue, coordination, coherence and synergies among stakeholders; and (iii) enhancing action and support including finance, technology and capacity building (UNFCCC, 2013a).

The WIM has an Executive Committee (ExCom) that meets approximately twice a year. The initial 2-year work plan of the WIM ExCom included nine activity areas. In 2017, at COP 23, the new 5-year workplan was approved. It included a smaller set of work streams that looked at slow onset events, non-economic losses, comprehensive risk management, migration and displacement, and action and support.

While the debate on climate change and Loss and Damage under the UNFCCC and the WIM has been largely political (Calliari et al., 2019), there is also a strong connection with the scientific community, and particularly with the IPCC. The IPCC plays an important role in the climate change negotiations as a provider of policy relevant information, involving government participation at different stages. Roberts and Huq (2015) show how important milestones in the climate negotiations have followed the presentation of more robust evidence on climate impacts and adaptation barriers in the assessment reports that the IPCC has published since 1990. The IPCC only assesses the existing literature on climate change, and does not conduct its own research. Therefore, knowledge gaps in the IPCC reports mostly reflect gaps in the literature.

The aim of this paper is to analyze how the terms 'loss' and 'damage' are used in IPCC WGII AR5. Through this analysis, the authors try to identify knowledge gaps in the report and areas that require attention from IPCC authors while they work on the Sixth Assessment Report (AR6), which is expected in 2021. The paper analyses which climatic stressors, impact sectors and regions WGII AR5 primarily associates with losses and damages, and tries to find out whether the report treats losses and damages primarily in connection to natural or human systems. For losses and damages to human systems, the paper looks at the relative attention given to economic and non-economic losses and damages.

The structure of the paper is as follows. First, we explain the methods used to analyse the more than 2500 pages[1] of the report (data mining with qualitative data analysis software). After that, the results and discussion section analyses the use of the terms loss(es) and damage(s) by chapter, and by studying the words used in one sentence with the terms loss(es) and damage(s) along four axes of thematic interest: type of climatic stressors, impact on natural and human systems, economic and non-economic losses and geographic region. The last section provides conclusions with implications for the WIM and AR6.

Materials and methods

Qualitative data analysis software (QDA Miner/WordStat) was used to extract sentences from the 30 IPCC WGII AR5 chapters plus the SPM and the Technical Summary (TS) containing the words loss(es), lost, losing, lose, loser (s), damage(s), damaged or damaging. The resulting 1,911 sentences were exported to a spreadsheet and screened for technical and formatting issues (e.g. incomplete sentences, more than one sentence, text in tables not correctly exported, illegible symbols, erroneous spaces, page breaks) and to check whether the words loss and damage were actually used in a meaningful way (e.g. author name: 'Scott R. Loss' was excluded).

The resulting document contained 1,886 sentences, in which loss, damage and related words occurred 2,177 times (in some sentences, the words occurred more than once). Losses were mentioned much more often than damages (see Table 1). Table 1 also compares the use of the words loss/damage in AR5 with the previous fourth

Table 1. Use of the words loss and damage in AR4 and AR5.

Key term	Frequency WGII AR4	Frequency WGII AR5
Loss	446	872
Losses	265	525
Damage	307	419
Damages	156	172
Lost	60	70
Damaging	19	42
Lose	27	23
Damaged	23	22
Loser(s)	7	18
Losing	8	14
Total	1313	2177

Source: Authors.

assessment report (AR4), published in 2007. It shows that both terms were used much more frequently in WGII AR5 than in AR4 WGII. The set of 1,886 sentences was first used for a simple analysis of the frequency of occurrence of the words loss/damage, followed by a more in-depth analysis of how losses and damages feature in the report.

In the second step, the file with 1,886 sentences was subjected to analysis to explore the words most often used in combination with loss/damage. A threshold was set at frequency 10, meaning that words that co-occurred with loss/damage less than 10 times were excluded from the analysis. The QDA software automatically excludes words that convey little intrinsic meaning, such as *about, above, according, across*, etc. The resulting list contained 587 words used in relation to loss/damage. This list was cleaned by:

- Removing author names;
- Removing words that conveyed no intrinsic meaning in this context, but were not automatically excluded by the QDA software (e.g. chapter, section, common, IPCC, SPM, terms, important, related, report, role, similarly, etc.);
- Clustering words with the same root (e.g. agriculture and agricultural). We were conservative in clustering words because sometimes words with the same root have a different meaning (e.g. effects and effective were kept separate, and so were developing and developed). In case of doubt, the original text was consulted to verify whether words conveyed the same meaning.
- In a few instances, words were combined (e.g. the word Zealand only occurred in New Zealand; sheet only in ice sheet, greenhouse only in greenhouse gas, etc.). When the other word (e.g. ice in ice sheet, sea in sea level rise) also occurred independently, the frequency score was adjusted (i.e. frequency of ice sheet deducted from frequency of ice).

The cleaned word list contained 301 words that occurred at least 10 times in the same sentence with the words loss(es) or damage(s). This list and the frequencies with which the words occurred was used to support the analysis of how WGII AR5 covers current and future losses and damages associated with impacts of climate change.

Limitations

The approach in this paper has several limitations. First, it covers only the contribution of WGII. The reason to limit the scope was made because the contributions of Working Groups I and III focus on the causes of climate change and options for reducing GHG emissions respectively, and not on the impacts of climate change. The use of QDA software to count frequencies with which the terms loss and damage appear in chapters and to analyse which words are used most frequently in combination with these terms, proved an effective method for analysing the more than 2,500 pages of the report. However, and this is the second limitation, the results of this analysis do not necessarily provide a full understanding of what WGII AR5 has to say about loss and damage. To address this limitation, the original text was frequently consulted to be able to provide background, interpretation and a more profound and qualitative understanding to the more quantitative findings. A third limitation of the QDA analysis is that it does not capture all instances in which WGII AR5 discusses evidence on losses and damages. For instance, the report often writes about the adverse consequences of climate change that are 'beyond adaptation' (Botzen et al., 2019) as regular climatic impacts without using the words 'loss' or 'damage'.

Results and discussion

Loss and damage by WGII AR5 chapter

In this section, we look at how often different chapters use the words loss and damage (see Figure 1). The analysis is a simple frequency score, distinguishing loss (including related words, such as losses, lost and losing) and damage (including related words, such as damages, damaging and damaged).

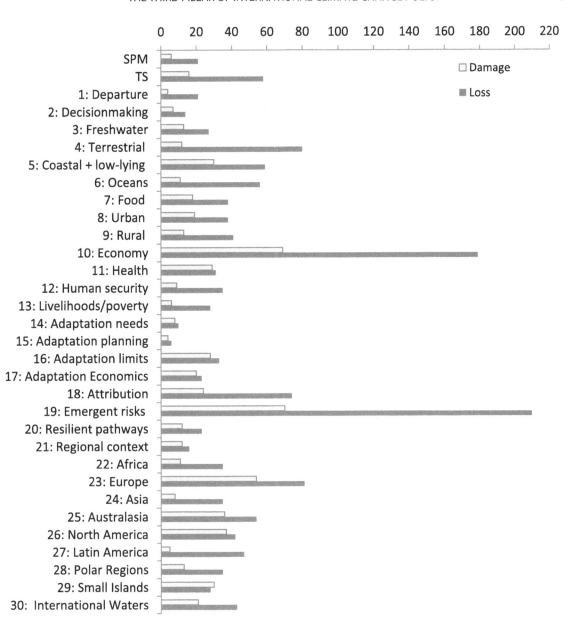

Figure 1. Occurrence of the words loss and damage by chapter. Source: Authors. The words included in the analysis are loss(es), lost, loser(s), losing, damage(s), damaged or damaging.

A first observation from Figure 1 is that the word 'loss' is used much more often than the word 'damage'. While some studies (e.g. Doelle & Seck, 2019; McNamara & Jackson, 2019; Tschakert et al., 2019) assign specific meanings to the words 'loss' (permanent impacts) and 'damage' (reversible impacts), it would be inaccurate to conclude that the adverse effects of climate-related stressors, reported in WGII AR5, tend to be irreversible. For example, when adverse effects of climate change on livelihood security are discussed, the authors usually speak of 'loss of livelihood' without implying that livelihoods are lost forever.

The words loss and damage are most frequently used in Chapter 19 (Emergent risks and key vulnerabilities) and Chapter 10 (Key economic sectors and services). This is an indication that losses and damages are mostly framed in economic terms and that they are primarily seen as future threats. While Chapter 10 states that the influence of climate change on the global economy is relatively small compared to other drivers, it does highlight more severe impacts in some countries. It states: 'Climate could be one of the causes why some countries are trapped in poverty, and climate change may make it harder to escape poverty' (IPCC, 2014, p. 663) According to Chapter 19, a severe risk of climate change for human systems is the loss of ecosystem services, which is often exacerbated by local human activities, including mitigation action, such as the production of bioenergy crops. According to Chapter 19, 'the risk of severe harm and loss due to climate change-related hazards and various vulnerabilities is particularly high in large urban and rural areas in low-lying coastal zones' (IPCC, 2014, p. 1042). Such areas are exposed to multiple hazards, such as sea level rise, storm surge, coastal erosion, saline intrusion and flooding. Key risks identified in Chapter 19 include food insecurity, loss of rural livelihoods caused by water scarcity and loss of coastal livelihoods due to sea level rise and acidification.

The lowest frequencies are in Chapter 15 (Adaptation planning and implementation), Chapter 14 (Adaptation needs and options) and Chapter 2 (Foundations for decision making). This is an indication that policy to address loss and damage is still in its infancy. A key message of Chapter 15 is that adaptation planning is improving but more complex than many assume. The chapter highlights an important gap in adaptation planning, namely that monitoring and evaluation of adaptation plans is inadequate, and that this needs to be systematized to know what actions are most efficient to reduce future losses and damages. Another key message of Chapter 15 involves the need to remove institutional barriers to effective adaptation planning. Chapter 14 notes a gap between adaptation needs and options to meet those needs – the adaptation deficit – and sees a role for 'procedures to deal with loss and damage' to fill this gap (Chapter 14, p. 845).

The terms loss and damage are used more often in the chapters on Europe, North America and Australia than in chapters on Asia, Africa, Latin America and Small Islands. This is surprising, because losses and damages are mostly associated with vulnerable countries such as small island developing states (SIDS) and least developed countries (LDCs).

Another observation from Figure 1 is that the words loss and damage are used substantially less often in the SPM (27 times in 44 pages) than in the Technical Summary (74 times in 76 pages). An explanation could be that the SPM needs to be approved line by line by member country governments, and that industrialized countries successfully minimized the use of the term, fearing that the rise of the concept would open the door to compensation claims. A summary of the approval session (38th session of the IPCC, 25–29 March 2014) shows that attempts by vulnerable countries to include loss and damage language in the text were resisted by industrialized countries (IISD, 2014).

Terms associated with loss and damage: an analysis along 4 axes

In this section, we look at the words used in combination with the terms loss and damage. First, all words are taken together, and illustrated visually in a tag cloud (see Figure 2). After that, the words are analysed along four axes: type of climatic stressors, impact on natural and human systems, economic and non-economic losses and geographic region.

Figure 2 shows a tag cloud of words that co-occurred at least 25 times in the same sentence with the words loss or damage in the 30 chapters plus the SPM and TS. The larger the word size, the more often mentioned in relation to loss/damage.

The word most often used in connection to loss and damage is 'risk' (383 times). This is an indication that the report talks about losses and damages mostly – but not exclusively – as a *future* threat. This is also in line with the analysis in the previous section, which showed that the words loss and damage most often occur in Chapter 19 which focuses on *emergent* risks and key vulnerabilities.

Other words used at least 100 times in combination with loss or damage were – in descending order – economic, impacts, flood, coastal, adaptation, ecosystems, events, species, insurance, water, sea, ice, costs, coral, infrastructure, biodiversity and land. The word 'events' is mostly used in 'extreme weather events'. By contrast, 'slow-onset events' were mentioned only once in relation to loss and damage.

The words in Figure 2 refer to climatic stressors, impact types, processes and potential solutions. Below, they are analysed in more detail, and in relation to other words. The use of the word 'adaptation' in one sentence with loss/damage is particularly frequent in Chapter 16. The central argument in that chapter is that there are limits

Figure 2. Tag cloud – Words used in one sentence with loss or damage. Source: Authors.

Notes: The threshold for inclusion in the figure was set at 25.

and constraints to adaptation (see also Dow et al., 2013) and that not all climate-related losses and damages can be avoided, even if mitigation and adaptation efforts are intensified.

Type of climatic stressor

Figure 3 shows the climatic stressors that are mentioned in WGII AR5 at least 10 times in one sentence with loss or damage. Floods clearly stand out as the climate-related stressor that is most frequently associated with losses

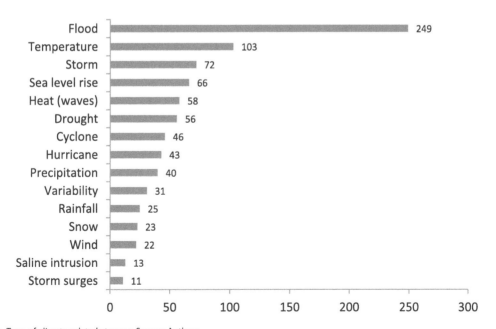

Figure 3. Type of climate-related stressor. Source: Authors.

Note: The threshold for inclusion in the figure was set at 10.

and damages. Second comes temperature (rise), which primarily causes losses and damages to ecosystems and animal and plant species. WGII AR5 discusses evidence of losses and damages from extreme weather events, such as floods, storms, heatwaves and cyclones/hurricanes, more frequently than evidence of losses and damages from more gradual and slow-onset processes, such as sea level rise and changing rainfall patters. While there is a long tradition of documenting losses from sudden-onset disasters, and these are well-documented in WGII AR5, the report cited much less work on losses and damages from incremental climatic changes.

Interestingly, from a climate science perspective, it is less complicated to attribute losses and damages to anthropogenic global warming in the case of slow-onset processes than in the case of extreme weather events (James et al., 2014). However, assessing losses and damages from those slow-onset processes tends to be more difficult (James et al., 2019). A complicating factor is that slow-onset processes and sudden-onset events usually interact. For example, sea level rise (slow-onset) exacerbates impacts of cyclones and tidal floods (sudden-onset). Also, sudden-onset events can act as triggers to push slower-onset changes over tipping points (van der Geest & Schindler, 2017). For example, a severe drought can trigger desertification. Another complicating factor in assessing losses and damages from slow-onset changes is that human systems have more time to adapt to these changes. Whereas an assessment of losses and damages from a cyclone would typically take place at a discrete point in time – usually soon after the cyclone – the timing of an assessment of losses and damages from sea level rise is less obvious.

Impacts on natural and human systems

Figure 4 shows words used in the same sentence as loss/damage that involve impacts on natural (the light blue bars) and human (the dark blue bars) systems. In some cases, a word can imply human impacts as well

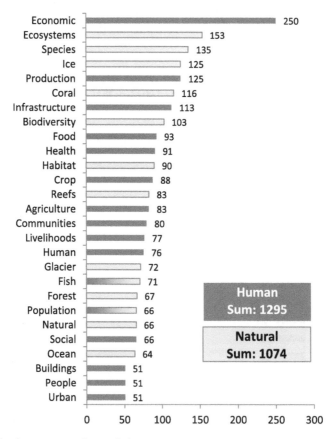

Figure 4. Impacts on natural vs human systems. Source: Authors.

Notes: The threshold for inclusion in the figure was set at 50.

as ecosystem impacts.[2] In such instances, the original text was consulted. The words fish and population were used in connection to impacts on human as well as natural systems. Other words, such as production, indigenous and diseases could in theory be used in both realms, but in practice were only used in relation to human impacts.

Overall, WGII AR5 pays a similar level of attention to losses and damages from climate change in human and natural systems (see Figure 4). In natural systems, the report expresses particular concern about losses and damages to ecosystems, species, habitat and biodiversity. Figure 4 also reveals an emphasis on marine and arctic ecosystems and less attention for terrestrial ecosystems. The impacts on human systems discussed in WGII AR5 primarily involve economic losses and damage to infrastructure. Substantially less attention is given to impacts on food security, health, livelihoods and communities, as is also shown in the next figure (Figure 5).

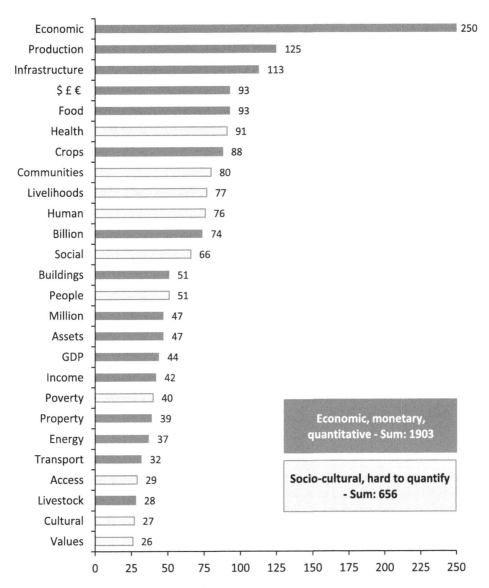

Figure 5. Economic versus socio-cultural losses. Source: Authors.

Note: The threshold for inclusion in the figure was set at 25.

Economic and socio-cultural loss and damage

Figure 5 plots words used in the same sentence with loss/damage that are related to climate impacts on human systems. A distinction is made between economic (the dark blue bars) and socio-cultural (the light blue bars) impacts, or to be more precise, between economic, physical, monetary and quantifiable impacts on the one hand, and socio-cultural, people-centred and hard to quantify impacts on the other. While a distinction is usually made between economic and non-economic losses, we feel that these labels are not fully adequate in this context. Economic losses are defined by the UNFCCC as 'losses of resources, goods and services that are commonly traded in markets.' (UNFCCC, 2013b, p. 3). By contrast, socio-cultural losses are understood in the technical paper as losses to things that are not commonly traded in markets, and therefore challenging to assess.

WGII AR5 reports losses and damages primarily in relation to physical, economic and monetary impacts (the dark blue bars) of climate change and extreme weather events (see Figure 5). On the people-centred side (the light blue bars), food security, health and livelihoods are the key sectors where climate change-related losses and damages are observed and expected. Climate change-induced food security problems are particularly expected in Sub-Saharan Africa, where temperature increases in some areas may be beyond adaptation limits, and where an increased frequency and intensity of droughts and floods would affect yield levels and post-harvest losses. Expected impacts of climate change on human health would result partly from food insecurity, but also from an increase in vector – and water-borne diseases, associated with global warming.

Below, we have listed a selection of quotes from different chapters that show that WGII AR5, despite its tendency to focus primarily on economic losses, also has some important things to say about non-economic, non-monetary, social and cultural losses and damages, such as displacement, loss of social identity and loss of damage to cultural heritage:

- SPM, p. 19: 'Disaster loss estimates are lower-bound estimates because many impacts, such as loss of human lives, cultural heritage, and ecosystem services, are difficult to value and monetize, and thus they are poorly reflected in estimates of losses.'
- TS, p.73: 'Loss of land and displacement, for example, on small islands and coastal communities, have well documented negative cultural and well-being impacts.'
- Chapter 5, p.364: 'Without adaptation, hundreds of millions of people will be affected by coastal flooding and will be displaced due to land loss by the year 2100; the majority of those affected are from East, Southeast and South Asia (high confidence).'
- Chapter 16, p.922: 'Strategies such as migration (…) may involve the loss of sense of place and cultural identity, particularly if migration is involuntary.'
- Chapter 29, p. 1639: 'Relocation and displacement are frequently cited as outcomes of sea-level rise, salinization and land loss on islands.'
- Chapter 23, p.5: 'Climate change and sea level rise may damage European cultural heritage, including buildings, local industries, landscapes, archaeological sites, and iconic places [medium confidence].' Geographic regions: continents, countries, regions

Regions

Whereas vulnerable countries, such as SIDS and LDCs were the main driving force behind the establishment of the WIM (Roberts & Huq, 2015; Calliari et al., 2019), surprisingly, WGII AR5 mentions *developed* countries much more often in relation to loss and damage. The words Europe, Australia, North America and United States co-occur with loss/damage about three times more often than the words Asia, Africa, Latin America and the Pacific (see Figure 6). Similarly, Germany is mentioned in connection to loss/damage more often than the entire Caribbean and almost twice as often as an extremely vulnerable country like Bangladesh. An explanation may be that more research has been done and more robust evidence was available in high-income countries (Hansen & Cramer, 2015). The composition of IPCC WGII author teams[3] might also play a role as authors from developed countries dominate (Ford, Vanderbilt, & Berrang-Ford, 2012) and may be more familiar with

Figure 6. Geographic regions: continents, countries, regions. Source: Authors.

Notes: The threshold for inclusion in the figure was set at 10. Latin America was not mentioned often enough in connection to loss/damage to be included in Figure 6. The authors acknowledge that Asia includes countries, such as Japan, South Korea and Singapore, that are considered developed countries.

evidence from their own regions. Another reason could be that economic losses, when expressed in monetary terms, tend to be higher in high-income countries. Examples from the United States are Hurricane Katrina and Super-storm Sandy, with an estimated economic damage of US$ 100 and 50 billion, respectively (Chapter 5, p. 383).

Conclusion

This paper used qualitative data analysis software (text mining) to study what WGII AR5 has to say about losses and damages from climate change. The words 'loss' and 'damage' occur over 2,000 times and we assessed which climatic stressors, impact sectors and regions the report primarily associates with losses and damages. In these concluding paragraphs we summarize key findings and highlight the implications for the WIM and the IPCC.

As a concept, 'Loss and Damage' does not feature prominently in WGII AR5, but the SPM and the TS do state with very high confidence that there is a risk of unavoidable losses and damages, despite current and future mitigation and adaptation efforts. Moreover, throughout the 30 chapters of the report, evidence of current losses and damages are presented, and the risks of future losses and damages are assessed. A clear message of the report is that postponing ambitious mitigation action increases the chances of crossing adaptation limits, and could lead to irreversible losses to ecosystems and society, particularly in low-income countries.

The word most often used in connection to losses and damages is *risk* (386 times) and the chapters in which the words loss and damage appear most frequently are chapter 19 (Emergent risks and key vulnerabilities) and 10 (Key economic sectors and services). This is an indication that the report talks about losses and damages mostly in economic terms and as a future threat. Non-economic losses and damages and the social and cultural dimensions of loss receive less attention. Furthermore, WGII AR5 does not include enough evidence about loss and damage as a reality for vulnerable people today. It is not entirely clear to what extent this is because there is a lack of evidence in the academic literature or because of the composition of the IPCC WGII author teams, which are dominated by economists and others from developed countries (Ford et al., 2012; Carey, James, & Fuller, 2014). A recommendation of this paper to the IPCC is to include more authors who are familiar with qualitative research on social and cultural dimension of climate change and loss and damage, especially in vulnerable countries. Typically such authors would hail from anthropology, development studies, human geography and psychology. A key resource for the IPCC, and its AR6, could be the expert group on non-economic losses that was established under the current five-year workplan of the WIM (Serdeczny, 2019).

IPCC WGII AR5 discusses losses and damages mostly in relation to floods and other extreme events, such as storms and hurricanes. It has less to say about losses and damages from incremental processes and gradual climatic changes. Here, too, it is not entirely clear to what extent this is because there is a lack of evidence in the academic literature or because IPCC WGII authors are less familiar with the evidence on losses and damage from incremental and slow-onset processes. While existing disaster loss databases and institutional structures for disaster management can play an important role in assessing and addressing losses and damages (Gall, 2015), the risks of losses and damages from slow-onset processes and gradual climatic changes, and the dangerous interaction between slow-onset processes and sudden-onset events (James et al., 2019), need more attention in IPCC AR6. Just as in the case of non-economic losses, a key resource for the IPCC, and its AR6, could be the technical expert group on slow-onset events that was established under the WIM.

Whereas vulnerable developing countries were the main driving force behind the establishment of the WIM, WGII AR5 mentions developed countries much more often in relation to losses and damages. However, the real losses and damages from climate change in terms of human suffering, disrupted livelihoods and undermined sustainable development pathways are particularly severe in the world's LDCs and SIDS. This is well-recognized under the WIM as its primary focus has from the beginning been on 'developing countries that are particularly vulnerable to the adverse effects of climate change' (UNFCCC, 2013a). The recommendation to the IPCC is to continue and intensify efforts to include more authors from developing countries, and particularly authors from LDCs and SIDS.

The chapters that are most policy-relevant[4] are also the most silent about loss and damage. This is not surprising because when WGII AR5 was prepared, climate policy focused almost exclusively on mitigation and adaptation. The Paris Agreement has the potential to change that, as it acknowledges that some losses and damages cannot be avoided through mitigation and adaptation policy. Separate policy is needed for such residual loss and damages.

Notes

1. We analysed the thirty chapters of IPCC WGII AR5 plus the SPM and TS, totalling 2605 pages.
2. The authors acknowledge that impacts on natural systems often affect human systems through loss of ecosystem services (Costanza et al., 1997; Zommers et al., 2016; van der Geest et al., 2019).
3. The IPCC website documents WGII author team composition. It shows that 41% come from developing countries or economies in transitions.
4. Chapter 15 (Adaptation planning and implementation), Chapter 14 (Adaptation needs and options) and Chapter 2 (Foundations for decision making).

Disclosure statement

No potential conflict of interest was reported by the authors.

References

Botzen, W. W., Bouwer, L. M., Scussolini, P., Kuik, O., Haasnoot, M., Lawrence, J., & Aerts, J. C. (2019). Integrated disaster risk management and adaptation. In R. Mechler, L. Bouwer, T. Schinko, S. Surminski, & J. Linnerooth-Bayer (Eds.), *Loss and damage from climate change* (pp. 287–315). Cham: Springer.

Bouwer, L. M. (2011). Have disaster losses increased due to anthropogenic climate change? *Bulletin of the American Meteorological Society*, *92*(1), 39–46.

Burton, I. (2009). Climate change and the adaptation deficit. In E. L. F. Schipper, & I. Burton (Eds.), *Earthscan Reader on adaptation to climate change* (pp. 89–95). London: Earthscan.

Byrnes, R., & Surminski, S. (2019). *Addressing the impacts of climate change through an effective warsaw international mechanism on loss and damage: Submission to the second review of the warsaw international mechanism on loss and damage under the UNFCCC.* London: Grantham Research Institute and London School of Economics and Political Science.

Calliari, E., Surminski, S., & Mysiak, J. (2019). The politics of (and behind) the UNFCCC's loss and damage mechanism. In R. Mechler, L. Bouwer, T. Schinko, S. Surminski, & J. Linnerooth-Bayer (Eds.), *Loss and damage from climate change* (pp. 155–178). Cham: Springer.

Carey, M., James, L. C., & Fuller, H. A. (2014). A new social contract for the IPCC. *Nature Climate Change, 4*(12), 1038.

Costanza, R., d'Arge, R., de Groot, R., Farber, S., Grasso, M., Hannon, B., … van den Belt, M. (1997). The value of the world's ecosystem services and natural capital. *Nature, 387*, 253–260.

Doelle, M., & Seck, S. (2019). Loss & damage from climate change: from concept to remedy? *Climate Policy.* doi:10.1080/14693062.2019.1630353

Dow, K., Berkhout, F., Preston, B. L., Klein, R. J., Midgley, G., & Shaw, M. R. (2013). Limits to adaptation. *Nature Climate Change, 3*(4), 305–307.

Fankhauser, S., Dietz, S., & Gradwell, P. (2014). Non-economic losses in the context of the UNFCCC work programme on loss and damage. In *Policy paper.* Centre for Climate Change Economics and Policy Grantham Research Institute on Climate Change and the Environment. Retrieved from http://www.lse.ac.uk/GranthamInstitute/wp-content/uploads/2014/02/Fankhauser-Dietz-Gradwell-Loss-Damage-final.pdf

Ford, J. D., Vanderbilt, W., & Berrang-Ford, L. (2012). Authorship in IPCC AR5 and its implications for content: Climate change and indigenous populations in WGII. *Climatic Change, 113*(2), 201–213.

Gall, M. (2015). The suitability of disaster loss databases to measure loss and damage from climate change. *Int J Global Warming, 8*(2), 170–190.

Hansen, G., & Cramer, W. (2015). Global distribution of observed climate change impacts. *Nature Climate Change, 5*(3), 182–185.

Huggel, C., Stone, D., Auffhammer, M., & Hansen, G. (2013). Loss and damage attribution. *Nature Climate Change, 3*(8), 694–696.

Hulme, M. (2014). Attributing weather extremes to 'climate change': A review. *Progress in Physical Geography: Earth and Environment, 38*(August), 499–511.

Huq, S., Roberts, E., & Fenton, A. (2013). Loss and damage. *Nature Climate Change, 3*(11), 947–949.

IISD. (2014). Summary of the 10th session of Working Group II of the Intergovernmental Panel on climate change (IPCC) and thirty-eighth session of the IPCC: 25-29 March 2014. *Earth Negotiations Bulletin, 12*(596), 1–20.

IPCC. (2007). *Climate change 2007: Impacts, adaptation and vulnerability. Contribution of working group II to the fourth Assessment report of the intergovernmental panel on climate change.* Cambridge: Cambridge University Press.

IPCC. (2014). *Climate change 2014: Impacts, adaptation and vulnerability. contribution of working group II to the fifth assessment report of the intergovernmental panel on climate change.* Cambridge: Cambridge University Press.

James, R. A., Jones, R. G., Boyd, E., Young, H. R., Otto, F. E., Huggel, C., & Fuglestvedt, J. S. (2019). Attribution: How is it relevant for loss and damage policy and practice? In R. Mechler, L. Bouwer, T. Schinko, S. Surminski, & J. Linnerooth-Bayer (Eds.), *Loss and damage from climate change* (pp. 113–154). Cham: Springer.

James, R., Otto, F., Parker, H., Boyd, E., Cornforth, R., Mitchell, D., & Allen, M. (2014). Characterizing loss and damage from climate change. *Nature Climate Change, 4*(11), 938–939.

McNamara, K. E., & Jackson, G. (2019). Loss and damage: A review of the literature and directions for future research. *Wiley Interdisciplinary Reviews: Climate Change, 10*(2), e564.

Mechler, R., Calliari, E., Bouwer, L. M., Schinko, T., Surminski, S., Linnerooth-Bayer, J., … Zommers, Z. (2019). Science for loss and damage. Findings and propositions. In R. Mechler, L. Bouwer, T. Schinko, S. Surminski, & J. Linnerooth-Bayer (Eds.), *Loss and damage from climate change* (pp. 3–37). Cham: Springer.

Meinshausen, M., Meinshausen, N., Hare, W., Raper, S. C. B., Frieler, K., Knutti, R., … Allen, M. R. (2009). Greenhouse-gas emission targets for limiting global warming to 2°C. *Nature, 458*, 1158–1162.

Nalau, J., & Leal Filho, W. (2018). Introduction: Limits to adaptation. In W. Leal Filho & J. Nalau (Eds.), *Limits to climate change adaptation* (pp. 1–8). Cham: Springer.

Preston, B., Dow, K., & Berkhout, F. (2013). The climate adaptation frontier. *Sustainability, 5*(3), 1011–1035.

Roberts, E., & Huq, S. (2015). Coming full circle: The history of loss and damage under the UNFCCC. *International Journal of Global Warming, 8*(2), 141–157.

Roberts, E., & Pelling, M. (2019). Loss and damage: An opportunity to address the root causes of vulnerability through transformational change? *Climate Policy.* doi:10.1080/14693062.2019.1680336.

Roberts, E., van der Geest, K., Warner, K., & Andrei, S. (2014). Loss and damage: When adaptation is not enough. *Environmental Development, 11*(July), 219–227.

Schellnhuber, H. J., Hare, B., Serdeczny, O., Adams, S., Coumou, D., Frieler, K., … Warszawski, L. (2012). *Turn down the heat: Why a 4 C warmer world must be avoided.* Washington, DC: World Bank.

Serdeczny, O. (2019). Non-economic loss and damage and the Warsaw international mechanism. In R. Mechler, L. Bouwer, T. Schinko, S. Surminski, & J. Linnerooth-Bayer (Eds.), *Loss and damage from climate change* (pp. 205–220). Cham: Springer.

Tschakert, P., Ellis, N. R., Anderson, C., Kelly, A., & Obeng, J. (2019). One thousand ways to experience loss: A systematic analysis of climate-related intangible harm from around the world. *Global Environmental Change, 55*, 58–72.

UNFCCC. (2013a). Report of the conference of the parties on its eighteenth session, held in Doha from 26 to 8 December 2012. Addendum. Part two: Action taken by the Conference of the Parties at its eighteenth session. United Nations Framework Convention on Climate Change (UNFCCC). FCCC/CP/2012/8/Add.1.

UNFCCC. (2013b). Non-economic losses in the context of the work programme on loss and damage. UNFCCC Technical Paper TP/2013/2.

van der Geest, K., de Sherbinin, A., Kienberger, S., Zommers, Z., Sitati, A., Roberts, E., & James, R. (2019). The impacts of climate change on ecosystem services and resulting losses and damages to people and society. In R. Mechler, L. Bouwer, T. Schinko, S. Surminski, & J. Linnerooth-Bayer (Eds.), *Loss and damage from climate change* (pp. 221–236). Cham: Springer.

van der Geest, K., & Schindler, M. (2017). *Handbook for assessing loss and damage in vulnerable communities.* Bonn: UNU-EHS.

van der Geest, K., & Warner, K. (2015). Loss and damage from climate change: Emerging perspectives. *International Journal of Global Warming, 8*(2), 133–140.

van Vliet, J., van der Berg, M., Schaeffer, M., van Vuuren, D. P., den Elzen, M., Hof, A. F., … Meinshausen, M. (2012). Copenhagen accord pledges imply higher costs for staying below 2°C warming. *Climatic Change, 113*(2), 551–561.

Verheyen, R., & Roderick, P. (2008). *Beyond adaptation: The legal duty to pay compensation for climate change damage.* Surrey: WWF-UK.

Warner, K., & van der Geest, K. (2013). Loss and damage from climate change: Local-level evidence from nine vulnerable countries. *International Journal of Global Warming, 5*(4), 367–386.

Warner, K., van der Geest, K., & Kreft, S. (2013). *Pushed to the limits: Evidence of climate change-related loss and damage when people face constraints and limits to adaptation.* Report No.11. Bonn: United Nations University Institute for Environment and Human Security (UNU-EHS).

Zommers, Z., van der Geest, K., De Sherbinin, A., Kienberger, S., Roberts, E., Harootunian, G., … James, R. (2016). *Loss and damage: The role of ecosystem services.* Nairobi: UNEP.

Loss & damage from climate change: from concept to remedy?

Meinhard Doelle and Sara Seck

ABSTRACT

In this article we examine legal perspectives on remedies for harm caused by climate related loss and damage. We start by discussing the meaning of loss and damage, and its relationship to climate mitigation and adaptation. We then consider, at a conceptual level, how those harmed by loss and damage from human-induced climate change may pursue remedies against those who have contributed to the harm suffered.

Key policy insights
- Loss and damage is an issue that requires the attention of law and policy makers at domestic and international levels
- While existing legal systems are unlikely to be adequately equipped in their present form to respond adequately to claims for remedy to harm caused by loss and damage, they will be challenged to evolve over time to respond more effectively
- Legal systems will be challenged to identify appropriate claimants, appropriate respondents, appropriate remedies and actionable wrongs
- Different legal systems will make different choices on these critical issues.

Introduction

Efforts to deal effectively with loss and damage (L&D) in the UN climate regime, and to provide for avenues to remedy associated harms, have so far failed (Lees, 2017; Siegele, 2017). While these efforts are ongoing, it is becoming increasingly clear that a broad range of international regimes and domestic legal systems will be challenged to respond to calls for appropriate remedies for those harmed by L&D. In this paper, we explore, at a conceptual level, the many issues that will arise as legal systems around the world are confronted with L&D claims. To provide some context for this analysis, we start in this introduction with a high-level overview of issues related to the scope and definition of L&D.

L&D is not defined in the UN climate regime. It has been suggested in the literature, however, that the phrase 'loss and damage' recognizes two categories of harm. One category involves permanent harm, or irrecoverable 'loss', such as the loss of landmass from sea level rise. The second category involves reparable or recoverable 'damage', such as shoreline damage from storms (CDKN et al., 2012; Morrissey & Oliver-Smith, 2013; Nishat, Mukherjee, Roberts, & Hasemann, 2013). Other ways the concept of L&D has been delineated is between economic and non-economic L&D, and between slow onset and extreme weather events (Fankhauser, Dietz, & Gradwell, 2014; Stabinsky & Hoffmaister, 2012). The focus has been on harm caused by human-induced climate change itself. A more controversial category of harm associated with climate change not clearly falling within the definition of L&D is harm caused by response measures, including by mitigation efforts, adaptation, and geoengineering.

A set of concepts that may help to further clarify the meaning and scope of L&D are the terms 'avoided', 'unavoided', and 'unavoidable' L&D, introduced by Verheyen (2012, p. 6) in one of the earlier research reports on the issue. 'Avoided' refers to the climate impacts prevented by existing mitigation efforts. 'Avoidable' refers to impacts that can still be avoided through enhanced mitigation and through adaptation. 'Unavoidable' L&D

are impacts that are not preventable through future efforts. They are already inevitable as a result of past actions and cannot be avoided even with best efforts. Unavoidable L&D is also referred to as 'locked in'.

It is important to consider the relationship between mitigation, adaptation and L&D. It is well recognized that the level of mitigation affects the scale of L&D. The more ambitious our collective mitigation effort, the less future L&D we will suffer. The relationship between adaptation and L&D is similarly close, but more complex. Indeed, when the Warsaw International Mechanism for loss and damage associated with climate change impacts (WIM) was established, it was placed within the Cancun Adaptation Framework. The preamble of the decision by the Conference of the Parties (COP) establishing the WIM acknowledges that L&D arising from climate change 'includes, and in some cases involves more than, that which can be reduced by adaptation' (Lees, 2017; Siegele, 2017, p. 226; UN, 2014).

Adaptation efforts are critical to reducing the amount of L&D caused by climate change. Much can be done to assist those affected by climate change, both human and natural systems, to adapt. Adjustments to agricultural and forest management practices to deal with changes in temperature or precipitation patterns are among the many examples. Of course, not everyone affected by climate change has the necessary capacity, resources or other means to maximize adaptation opportunities. This means that there may be theoretical opportunities to avoid L&D through effective adaptation that are not realized (Van Den Homberg & McQuistan, 2019). This, in itself, makes it difficult to draw a clear line between adaptation and L&D.

The issue of displacement is illustrative of the complex inter-relationship between adaptation and L&D. If we take a hypothetical small island state that is unable to protect some or all of its territory from sea level rise, one might be inclined to view this as a failure of adaptation, and the resulting impact as L&D suffered by the residents of the affected small island state. However, the failure to protect its territory could either be as a result of technical adaptation limits, or it could be related to the lack of financial resources to implement the necessary measures. Furthermore, how the small island state itself (in case of internal displacement) and the global community (in case of external displacement) responds to the loss of territory will ultimately affect the scale and distribution of the resulting harm. How much say do those displaced have over the preferred solutions? To what extent do the solutions cause L&D to others adversely affected by these solutions? To what extent do the solutions offer opportunities either to those displaced or to those who receive them? Is the focus on individual impacts or on collective L&D, such as loss of culture and community? Are efforts to find solutions for displaced persons to minimize their individual or collective L&D considered adaptation, or is adaptation limited to efforts to preserve the territory of the small island state (Mayer, 2014; McNamara, Bronen, Nishara Fernando, & Klepp, 2018; UN, 2014)? These are among the issues that arise in efforts to understand and delineate the complex relationship between adaptation and L&D.

Efforts under the UN climate regime to fully integrate L&D into the finance, transparency and stocktake elements of the Paris Agreement have been met with strong resistance from key developed countries. This has important implications for the consideration of L&D in the future, particularly its role in the 5-year review cycles under the Paris Agreement designed to increase ambition over time. The future of the issue within the UN climate regime generally remains uncertain, and the prospects for addressing funding needs to actually address L&D remain bleak. The focus, for now, will continue to be on improving understanding of the challenge, and to explore non-monetary avenues to help Parties manage the impacts (Lees, 2017; Siegele, 2017; UN, 2018).

Attention to what avenues might exist outside the UN climate regime to pursue remedies for L&D leads to a wide range of conceptual questions that will be the primary focus of this paper. For example, within the climate regime, one might presume that the actors seeking remedies for L&D (if, indeed, remedies were to be available) would be states, although initial discussions tended to treat 'vulnerable countries as populations', rather than states (Mayer, 2014). Similarly, under the climate regime, it has been presumed by most Parties, at least for now, that those who might have a responsibility to fund L&D are also states (Gewirtzman et al., 2018; Siegele, 2017). Outside the climate regime, this clearly cannot be presumed.

As perspectives on L&D from outside the climate regime are considered, attention shifts to a wide range of actors and institutions, and new areas of law, that are all potentially relevant to the search for remedies. For example, what is the relevance, if any, of migration and refugee law, disaster law, law of the sea, or international human rights law, to the question of L&D for climate harms? What is the relationship between L&D, and climate justice? Many areas of law will be challenged to deal with L&D, and as a result, the issue needs to be considered from a great variety of perspectives.

Legal perspectives on the future of loss and damage

In this section, we consider L&D through the lens of potential legal and related strategies for those who have experienced climate harms. Specifically, we consider different ways to frame the harm suffered, potential actionable wrongs, remedies, and parties to a L&D dispute. We draw on experience to date from climate litigation and legal concepts drawn from other contexts such as insurance, which aims to spread risk while providing the insurer with the ability to sue a third party to recover costs paid out. However, the aim here is conceptual, in recognition of the fact that law must evolve to address climate L&D challenges, and domestic laws will differ from jurisdiction to jurisdiction. In many areas of law, it may be too early to predict the direction that litigants, courts, and law makers will take. The aim therefore is to explore options rather than to predict specific directions, approaches or outcomes in specific domestic or international legal systems. We have also consciously chosen not to adopt a particular theory to inform our exploration of L&D. However, as appropriate, we observe that some legal and related strategies will align more closely with climate justice or human rights-informed approaches to L&D, while others will suggest the potential of transnational or multi-level governance. Attention will also be paid to key foundational concepts such as climate vulnerability and adaptive capacity.

(1) Framing the Harm Suffered

Whether it is in the context of the WIM, insurance, funding mechanisms or liability, defining harm from L&D resulting from climate change will be critical. The approach taken to delineating harm from L&D will undoubtedly vary depending on the context, including whether the harm is reversible, whether it was avoidable, and perhaps whether it can be expressed in monetary terms. Moreover, courts faced with claims might reach a different conclusion on what constitutes L&D than an insurance or compensation scheme set up to protect farmers in a developing country from the risk of drought or flooding, for example.

One of the challenges will be to separate harm linked to anthropogenic climate change from other sources of the harm experienced. Extreme weather events such as wind storms will be amplified by climate change, but in many cases will not be solely attributable to it. Similarly, harm associated with heat waves, droughts, floods, coastal erosion, melting permafrost, melting ice, or warming oceans will be exacerbated, but may not always be solely caused, by climate change (IPCC, 2014, 2018). Health impacts may face similar challenges, depending on whether climate change introduces a whole new health risk (such as the introduction of a disease to a region) or exacerbates an existing risk (such as health impacts of heat waves) (WHO, 2018). Separating the impact of climate change on such harm will be among the challenges of dealing fairly with harm from L&D (Marjanac, Patton, & Thornton, 2017).

Harm related to displacement illustrates the challenge of separating harm linked to climate change from other sources of the harm. Consider a farmer in a developing country who is internally displaced after successive years of crop failure. The farmer's plight may have been significantly influenced by climate change-induced droughts, flooding, storms or other changes to the climate system, but they may have also been influenced by changes in markets for the crops grown, unavailability of manual labour, or financial mismanagement of the farm.

The challenge of separating climate related harm from other sources of harm is, of course, not limited to harm that arises directly from the impacts of climate change. Similar issues arise with respect to harm caused by response measures to climate change, such as geoengineering. If ocean fertilization is attempted, for example, to increase the carbon uptake of the world's oceans, such efforts have the potential to cause harm to those depending on ocean resources. A key challenge will be to decide which harms related to climate response measures to include under the concept of L&D, such as whether to include those affected by an inequitable transition or otherwise harmed by efforts to mitigate climate change.

A key potentially unifying concept in this regard will be how to separate L&D from baseline conditions. The details will vary, but in each case the harm will have to be assessed against what would have happened without anthropogenic climate change (Allen et al., 2007). The nature of the harm, whether slow onset or arising from an extreme weather event, will obviously be relevant to this assessment. Moreover, as part of this process, it will be crucial to address the temporal dimension when selecting the baseline conditions against which climate harms are to be measured: are harms to be measured from the beginning of the industrial revolution?

Arguments might be advanced that the baseline for measuring climate L&D should be the start of global efforts to reduce emissions in 1990, or the moment at which global temperature increases surpass the Paris Agreement's 1.5 (or well below 2) degree goal. A related consideration is whether future harm will be included in the assessment of harm. If so, projections of future harm will be significantly dependant on mitigation efforts. More generally, it is far from clear whether the legal systems challenged to deal with harm from L&D need to prepare for harm caused by a 1.5 degree world, or 2, or 3, or perhaps even 4 or 5 degrees Celsius (Climate Action Tracker; IPCC, 2014, 2018).

Another key consideration is whether L&D will include only those impacts that have economic consequences, or whether it will extend to non-economic impacts, including cultural harms to indigenous peoples, for example (Fankhauser et al., 2014; Serdeczny, Bauer, & Huq, 2018). Clearly, from an indigenous environmental justice perspective, it will be essential that cultural dimensions of harm are acknowledged and included in L&D claims (Tsosie, 2007; Watt-Cloutier, 2016). Other questions remain. For example, will claims for L&D include loss of state territory due to sea level rise or even loss of statehood (Rayfuse & Crawford, 2011)? Would L&D to natural systems be viewed as harms that warrant a remedy? Answering these questions might involve evaluating the benefits provided to humans by ecosystem services, including those which underpin human livelihoods, as well as the ethical considerations of human-induced harms to non-human species (Diaz et al., 2018; UNEP, 2016; Zommers, Wrathwall, & Can der Geest, 2014).

(2) Potential Claimants

Closely related to the question of harm is who should be able to claim a remedy. The question of who has experienced harm will often arise in the context of litigation, but can also surface in the context of insurance and funding mechanisms set up to compensate victims. Entities that may experience harm as a result of unavoided L&D include states, sub-national government actors, as well as a variety of non-state actors ranging from individuals to organizations and communities. Examples of non-state actors who may be disproportionately impacted by L&D include indigenous and non-indigenous communities, migrants and refugees, children, women, and other vulnerable members of societies. Harm is, of course, not limited to humans, but includes human property as well as nature, from individual vulnerable species to whole ecosystems.

Not every entity potentially harmed by unavoided L&D will necessarily be entitled to a remedy. Any institution or legal regime that is asked to deal with unavoided L&D will be challenged to determine who will be eligible to seek a remedy; questions of legal standing will thus become important (Stone, 2010). Where those eligible include children, future generations and non-humans, questions arise as to who will be able to seek a remedy on their behalf.

A related consideration is whether L&D should be conceptualized as private harms (to individuals or groups in their private capacity) or as public harms (to societies as a whole, including future generations, or to public goods like ecosystems), or both. The global and transnational scale of climate harms further suggests that insights as to the nature of claimants from both public and private international law will be relevant, alongside those from domestic legal systems.

For example, it is clear that states are harmed by climate change, with some states, notably small island developing states (SIDS), being particularly vulnerable. International tribunals, such as the International Court of Justice (ICJ), the International Tribunal for the Law of the Sea (ITLOS), and the Permanent Court of Arbitration may grant standing to states who may bring claims for violations of their rights under international law (Bodansky, 2017; Strauss, 2009; Voigt, 2016).

Alternatively, the concept of *erga omnes* obligations at international law suggests that claims could be brought by one or more states on behalf of the international community as a whole for harm to the climate system or areas beyond national jurisdiction (Brown & Seck, 2013; UN, 2001, art. 48). A public international law perspective that conceptualizes L&D as a crime would not obviously provide an opportunity for plaintif states to bring L&D claims to the International Criminal Court, although prosecution of climate crimes has garnered scholarly and activist attention (Centre for Climate Crime; Ecocide Law Expert; Gallmetzer, 2017; Jodoin & Saito, 2011). If conceptualized as crimes, domestic statutes that implement international criminal law raise a theoretical opportunity for domestic prosecution of climate crimes, or, where parallel civil liability regimes exist, for individuals and groups that have experienced climate harms to seek a civil remedy (Wanless, 2009).

Private international law also offers insights into the nature of potential plaintiffs in transnational civil liability actions. For example, in many jurisdictions, foreign plaintiffs may bring claims provided that there is a real and substantial connection to the jurisdiction in which the action is brought. The presence of the defendant in the jurisdiction would often be a presumptive connecting factor, although judges may have discretion to decline to exercise jurisdiction on the basis of *forum non conveniens* or similar doctrines (see generally Amnesty International, 2014; Byers, Franks, & Gage, 2017; Seck, 2000). Another procedural consideration of relevance to delineating the plaintiffs in L&D litigation is the use of class action certification, which allows an individual plaintiff to file an action on behalf of themselves and others who are similarly situated. This access to justice mechanism is already in use in youth climate litigation (e.g. Trudel, Johnston, & Lespérance). In the future, questions may arise as to whether or how to delineate a global class in a L&D case, as has been the case in the consumer protection context, for example, where all consumers who suffered the same harm may be able to recover from a single settlement agreement (e.g. Global Class Actions Exchange; Clopton, 2018).

A different issue may be whether particularly vulnerable claimants should be given priority to claim a remedy over claimants who are in a more privileged position, and if so, how this might be achieved. For example, if any vision of climate justice is to be taken seriously, access to justice for L&D for the most vulnerable must be a priority (Adelman, 2016; Humphreys, 2014). It appears problematic, then, that comparatively privileged claimants who reside in a developed country that is also the home state of a corporate fossil fuel defendant, should face fewer legal hurdles than those that would confront a more climate-vulnerable plaintiff who happens to reside in a least developed country where no comparable defendant company has sufficient presence to ground jurisdiction. This is particularly problematic as those who reside in a least developed or most vulnerable country context may also face a lack of governmental capacity to offer adequate remedies to those who suffer from L&D. Moreover, individuals within these vulnerable states may be those who have contributed least to the climate problem on a per capita emissions basis, may not have benefited from economic and social wellbeing built up over years of fossil fuel extraction, or may live in isolation from the economic engine that drives greenhouse gas (GHG) emissions, as is the situation with many indigenous communities. Ultimately, the question is whether everyone should be entitled to L&D remedies, or whether access to climate justice should only – or first – be available to vulnerable groups within developing countries, or perhaps to the south within the north (e.g. indigenous peoples). In an ideal world, it would be possible to adequately compensate all who suffer harm, or otherwise provide a remedy. In our very imperfect world, however, it is more likely that defendants will seek bankruptcy protection before full compensation has been paid out – if any has been paid at all (Benjamin, Doelle, Jackson, & Seck, 2019). Furthermore, is there a duty on claimants to take steps to mitigate climate harms? In cases where such a duty is found to exist, a follow-up consideration is the extent of such a duty. For example, would a person suffering harm from climate change have a duty to relocate? Under what conditions? Different legal systems can be expected to reach different conclusions on these difficult questions, yet a lack of coordination among jurisdictions may leave the most vulnerable without access to justice.

(3) Potential Remedies

Another challenge for legal institutions and regimes in responding to unavoided L&D is to consider appropriate remedies. It is clear from the work under the WIM that only some harm resulting from unavoided L&D can be adequately quantified in economic terms (UNFCCC, 2014). This will challenge the ability of legal institutions to offer appropriate remedies, and will result in increasing pressure to expand the range of remedies beyond monetary compensation. Already, we are seeing climate litigation seeking a range of remedies, including declarations and various forms of injunctions to prevent undesired, or mandate desired, action. The case of *Urgenda v The Netherlands* [*Urgenda*] (2015), where the court imposed a minimum emission reduction target on the government of the Netherlands, serves to illustrate what lies ahead. Legal systems will increasingly be challenged to consider a broad range of remedies, including restitution, satisfaction and other forms of reparation (Burkett, 2009).

Diverse legal approaches may lead to monetary damage awards. For example, fines imposed as a result of successful criminal or regulatory climate prosecutions could be deposited in environmental damages funds to be distributed to victims (Government of Canada, 2018). Where climate remedies are pursued through a civil action, it is possible that punitive damages could be awarded. For climate litigation involving a transnational dimension, it is more likely that a monetary damage award would be recognized and enforced by a foreign court than a non-monetary judgement (Amnesty International, 2014; Byers et al., 2017; Seck, 2000).

If the distinction between loss as irreversible and damage as reversible is accepted, it may be useful to consider, separately, appropriate remedies for each. Legal systems will be challenged to consider whether and how to monetize harms such as loss of ecosystem services, loss of species, and loss of opportunity for children and future generations (Collins & McLeod-Kilmurray, 2014; Lord, Goldberg, Rajamani, & Brunnée, 2012). An additional complication is whether funds or support would be available for professional evaluation as to the existence and nature of these types of harms, such as assessment of habitat loss and recovery for species at risk, as well as compensation for losses while habitat recovers (Desierto, 2018). Drawing upon resilience theorists and others, would funding or support for recovery from L&D necessarily mean returning people to a place from which they came, or rebuilding ecosystems to the state that they were 'before' (assuming a precise time can be identified)? Or would L&D be reconcilable with the transformation of socio-ecological systems from one state to another? Given the unique characteristics and constraints of the wide range of regimes challenged to deal with L&D, one must expect a range of approaches to these difficult questions. Legal systems that are able to devise effective remedies for L&D will have the potential to contribute to more fundamental societal transformations essential for resilient futures (Morrissey & Oliver-Smith, 2013; Wrathall, Oliver-Smith, Fekete, & Sakdapolrak, 2015).

Litigation is only one avenue through which to seek a remedy. Others include compensation funds and insurance mechanisms. For example, crop insurance can offer remedies to farmers who may still be able to continue farming, but who face some risk every growing season that their crop will be harmed or destroyed by climate-induced extreme weather events such as flooding, drought or storms. Another example could be that of SIDS, who might be best served through a combination of disaster risk reduction and management funds, risk transfer through insurance, and ultimately compensation and rehabilitation (Burkett, 2015).

A final consideration is the relationship between geoengineering and L&D. Might arguments be made that geoengineering to reduce GHGs in the atmosphere or oceans should be understood as the equivalent of clean-up of environmental contamination or remediation of environmental harm? This possibility forces us to ask who should get to decide (and how should decisions be made) about the appropriateness of steps to reduce L&D, and the harm that may result from such steps. In a traditional environmental enforcement action, measures taken to reduce the extent of environmental harm and to quickly clean up the harm, may inform whether or not a (potential) offender is prosecuted, whether or not they meet a due diligence or reasonable care defence, and whether or not they are subject to a harsh or lenient penalty in sentencing. How could these concepts be adapted to the problem of climate L&D?

(4) Potential Defendants and Sources of Funds

Just as there is a broad range of possible plaintiffs seeking a remedy for harm resulting from unavoided L&D, there are many possible defendants. To date, defendants in climate litigation have included governments, companies that have contributed significantly to GHG emissions, and companies that have actively hindered the development and implementation of effective climate policies (e.g. Adler, 2018; Ganguly, Setzer, & Heyvaert, 2018; *Leghari v Pakistan*, 2015). Who will be an appropriate defendant will vary depending on the harm, the legal system called upon to provide a remedy, and the conduct that contributed to the harm. States, state actors, state owned enterprises, international organizations, and private actors are all potential defendants depending on the nature of the harm, the plaintiff, the legal system involved, and the remedies sought. Identifying appropriate defendants will be among the critical issues to be resolved for any legal system challenged to deal with liability or compensation for L&D, one that is closely connected to the issue raised in the following section, the question of the actionable wrong.

From a public international law perspective, defendants in actions brought before international tribunals like the ICJ must be states or international organizations, although in specific circumstances private actors may be subject to the jurisdiction of the deep seabed chapter of ITLOS (Stephens, 2009, p. 44). International civil liability treaties in areas such as oil pollution at sea, or nuclear accidents, on the other hand, while not directly applicable, could in theory offer inspiration for climate L&D litigation against multinational enterprises, or models for the creation of compensation funds with levies payable by polluters (Brown & Seck, 2013; Lyster, 2015).

If claims were brought in international courts or tribunals alleging climate harms against perpetrator states, it would be necessary to identify which states to target. Would the most appropriate defendant states be determined based on historic emissions by state, or would per capita calculations be relevant? Would emissions trading under the climate regime be relevant to the determination of a state's cumulative GHG emissions? What relevance, if any, would be given to a state's nationally determined contribution (NDC) under the Paris Agreement? Would it matter if a state met its targets, and if the targets were adequate, a 'fair share' of emissions for a 1.5 degree world? Would arguments be made that emissions associated with the production of goods for export should be attributed to purchaser countries? Would all actions of the state since the start of global efforts to address climate change around 1990 be relevant?

Beyond this, while some states have accepted compulsory jurisdiction of the ICJ, for example, others have shielded themselves through reservations (see Strauss, 2009 for the challenges associated with climate litigation at the ICJ). Another question might be whether international organizations could be defendants in international climate actions brought in domestic or international tribunals: could it be argued that the World Bank has contributed to climate L&D through its continued support for fossil fuel development around the world? (see *Jam v International Finance Corp*, 2019 finding the IFC has limited immunity and the commercial activity exception applies).

Prosecutions of international criminal law are designed to be brought against individuals rather than states, and corporations remain beyond the reach of international criminal justice for now. Transnational civil litigation in the climate context could be brought against a wide range of defendants, although challenges might arise due to legal doctrines of separate corporate personality that favour entity over enterprise liability except in exceptional circumstances (e.g. Chambers & Vastardis, 2018; *Yaiguaje v Chevron Corporation*, 2018; but see *Vedanta Resources PLC v Lungowe*, 2019). A related challenge is how to bring claims in foreign courts against states or state-owned enterprises that might challenge the exercise of jurisdiction on the basis of the act of state doctrine, or comity of nations (*Araya v Neysun*, 2017; Seck, 2017, p. 404). Heede's 'carbon majors' research, which has been relied upon in the Philippines climate petition, recognizes that while the 'private' investor-owned carbon majors have contributed greatly to GHG emissions since the beginning of the industrial revolution, so have state-owned enterprises and nation states (Greenpeace, 2015; Heede, 2014; Seck, 2017).

Clearly, evidence of attribution of harm will be crucial to identifying possible defendants, but in the climate context, complexities abound. How should the responsibility of a carbon major be measured? Should responsibility (and liability) be measured purely on the basis of contributions to global emissions, or is it relevant where that carbon major is based and the 'carbon budget' that its home state might be equitably allocated (as its 'fair share') (Expert Group, 2018)? Would a defendant multinational enterprise be liable for all emissions arising from the enterprise as a whole, or would separate legal personality of corporate entities and the contractual nature of supply and value chain relationships inhibit legal responsibility?

Different issues might arise in the identification of those most appropriate to contribute to international funds, even as contribution does not equate with a finding of legal liability for climate L&D. For example, it has been suggested that a levy on international airline passengers or bunker fuel used in marine transport, or even a fossil fuel majors carbon levy, could serve to gather funds necessary for L&D (Some of these are explored in Durand et al., 2018). Of course, not all ideas put forward to finance or insure climate remedy and compensation require the identification of a defendant-like contributor. For example, suggestions have been made to impose a financial transaction tax to fund L&D (Durand et al., 2018) while other commonly proposed insurance and finance tools include risk pooling, risk transfer, catastrophe risk insurance, contingency finance, climate themed bonds, and catastrophe bonds (Durand et al., 2018). Concerns have been raised that some of these mechanisms would place the burden of purchasing coverage upon the most vulnerable, rather than making the polluter pay (Burkett, 2009).

(5) Actionable Wrongs

Closely connected to the selection of appropriate defendants (or persons potentially responsible for proving the remedy) is the issue of the actions, omissions, or actionable wrongs that would hold the defendant responsible for contributing to the remedy. At one end of the spectrum, the 'wrong' may simply be the GHG emissions a company or state is responsible for (Heede, 2014). A L&D fund or insurance scheme, for example, could be set up on a no-fault absolute liability basis, similar to the oil pollution funds set up to contribute to the clean-up of oils spills caused by tanker traffic (Lyster, 2015).

Actionable wrongs, however, can take a range of forms. Exxon, for example, is being sued in the US for misleading investors about the risk climate change poses to the company (*People of the State of New York v Exxon Mobil*, 2018). A range of actionable wrongs are conceivable, ranging from lobbying against climate action domestically or internationally to funding others to undermine climate action. Some non-governmental business organizations funded by the fossil fuel industry, for example, have had a reputation of working to undermine consensus and ambition at the UN climate negotiations for years. Actions to mislead the public or governments about the science of climate change similarly could qualify as actionable wrongs similar to litigation in the tobacco context (Olszynski, Mascher, & Doelle, 2018).

Actionable wrongs can also take the form of government action or inaction. The *Urgenda* case in the Netherlands and the *Juliana* case in the US are perhaps the most prominent examples to date, but actionable wrongs by state actors can range from the role they play in international negotiations to failing to set adequate domestic targets and failing to meet targets set. Potentially, the dismantling of effective climate policy implemented by a previous government could be an actionable wrong by a government, as could efforts to undermine global ambition through a range of possible actions or inactions within the UN climate regime or other international fora seeking to raise the ambition of global action (*Juliana v United States*, 2016; *Urgenda*, 2015). Indigenous peoples experiencing social and cultural loss due to climate change might also claim actionable wrongs against governments and other actors whose failure to reduce emissions is equivalent to a failure to protect and respect their rights to self-determination, or other rights such as free, prior and informed consent that are recognized in the UN Declaration on the Rights of Indigenous Peoples (United Nations, 2007). The consequence of these rights violations is to undermine social and cultural resilience, including the legal orders of the indigenous peoples themselves.

A difficult challenge will be to distinguish those actionable wrongs that arise in the climate context more generally from those that might be said to be specifically designed to secure a remedy for climate L&D. It has been suggested that states could implement climate liability statutes to facilitate actions brought in domestic courts against climate polluters, similar to the way in which tobacco liability statutes were imposed in certain developed countries to recoup state healthcare costs (Byers et al., 2017; International Bar Association, 2014; Olszynski et al., 2018). Domestic climate litigation brought by domestic plaintiffs have raised a multitude of causes of action against their own governments as well as against business actors. Yet to date, only a few examples of climate litigation integrate a transnational dimension which has implications for the framing of the actionable wrong due to questions of choice of law. For example, if a low-lying island state sues a multinational fossil fuel company in the company's home state courts, should the (tort?) law of the multinational's home jurisdiction be applied to determine whether or not the multinational committed a wrong, or should the state be able to rely instead on its own climate L&D statutory cause of action (with a reversed burden of proof and reduced evidentiary burden)?

Irrespective of the cause of action, climate litigation will need to grapple with how to allocate responsibility where multiple contributors to the harm may be identified and proof of causation is elusive. This raises the possibility of joint and several liability (where plaintiffs may recover in full against one defendant who may then seek to recover from other equally responsible defendants), or alternate theories of liability such as material contribution to risk (where multiple defendants contribute to the risk of harm but no single defendant can be proven to be the necessary cause), or market share liability (where responsibility is allocated according to the defendant's share of global emissions) (Byers et al., 2017; Collins & McLeod-Kilmurray, 2014). Other models that could account for the challenge of attributing weather events to anthropogenic emissions include the fraction attributable risk allocation concept, drawn from epidemiological approaches, (Allen et al., 2007) or reliance on probabilistic event attribution (PEA) science (Parker et al., 2017).

Even if a plaintiff successfully argues that a defendant has committed the act necessary to find liability, defences or mitigating factors may ultimately detract from this result. How would courts address arguments that plaintiffs contributed to their own harm due to a failure to adapt? What about arguments that the plaintiff lacked capacity to adapt, or that there was a failure (by someone else?) to offer resources and other means for the plaintiff to adapt? What about ineffective implementation of adaptation measures? In the case of corporate defendants, questions could arise as to whether the defendant has met a reasonable care or due diligence defence – if so, which corporate social responsibility or climate-related industry standards might be relevant to this determination? Would proof of the establishment of an internal carbon pricing policy be seen as a mitigating factor? What about evidence of a plan to transition to carbon neutrality (such as from coal mining to wind energy)? Could evidence of misleading regulators and the public by funding climate deniers serve as an aggravating factor for punitive damages or as a separate cause of action?

Different insights emerge if causes of action are considered from the perspective of public international law. For example, a claim rooted in state responsibility would need to identify an internationally wrongful act that would be attributable to the defendant state. This could include arguments that a state's failure to regulate non-state actors to prevent climate harms could be an independent wrong in itself. State financing and support of a wrongful act by another state could also be a wrong (Brown & Seck, 2013). A state liability approach might hold a state strictly liable for environmental harm (that is, causing the harm is the wrongful act, a breach of the 'do no harm' principle) (Brown & Seck, 2013; UN, 2006).

A key question in the L&D context is how it will be determined if a state has committed a wrongful act for the purpose of state responsibility, in light of the requirements of the Paris Agreement. Are core international environmental law principles (such as do no harm) still relevant, or are they subsumed by the commitments of the Paris Agreement? What is the relationship between existing international human rights obligations to protect and ensure the realization of rights, and the commitments in the Paris Agreement (Wewerinke-Singh, 2018)? What about other treaty sources including the UN Law of the Sea with its obligations to protect the marine environment (Ohdedar, 2016)? Further, whose interpretation of contested international law norms will determine the answer to these questions? As evident from the emergence of civil society human rights tribunals like the Permanent Peoples' Tribunal on Human Rights, Fracking, and Climate Change, some would argue that the answers to these important questions cannot rest with the states and international bodies that have failed to protect the rights of the people of the world (Permanent People's Tribunal, 2018).

Conclusion

Prospects for a comprehensive solution to L&D under the climate regime are tenuous at best. This means that many international and domestic legal systems will increasingly be challenged to deal with aspects of L&D and to offer a remedy to those harmed. In this article, we have considered, at a conceptual level, the range of issues that legal systems around the world will need to grapple with.

We have considered who might bring claims for L&D, what remedies might be sought, who the remedies might be sought against, and what the actionable wrong might be. These are issues that both domestic and international legal systems challenged to respond to claims for L&D will face. Of course, each will likely take its own unique approach, resulting in a patchwork of venues with a different mix of eligible claimants, respondents, remedies and actionable wrongs. It remains to be seen whether this patchwork will be able to eventually develop into a cohesive whole that offers appropriate remedies to all legitimate claimants.

Disclosure statement

No potential conflict of interest was reported by the authors.

References

Adelman, S. (2016). Climate justice, loss and damage and compensation for small island developing states. *Journal of Human Rights and the Environment, 7*(1), 32–53.

Adler, D. P. (2018). US climate change litigation in the age of Trump: Year one. Retrieved from Columbia Law School, Sabin Center for Climate Change Law http://columbiaclimatelaw.com/files/2018/02/Adler-2018-02-U.S.-Climate-Change-Litigation-in-the-Age-of-Trump-Year-One.pdf

Allen, M., Pall, P., Stone, D., Stott, P., Frame, D., Nozawa, S. T., & Yukimoto, S. (2007). Scientific challenges in the attribution of harm to human influence on climate. *University of Pennsylvania Law Review, 155*(6), 1353–1400.

Amnesty International. (2014). *Injustice incorporated: Corporate abuses and the human right to remedy*. London, UK: Peter Benenson House.

Araya v Neysun. (2017). BCCA 401, 419 DLR (4th) 631, on appeal to the SCC.

Benjamin, L., Doelle, M., Jackson, C., & Seck, S. (2019, Feburary 22). Reflections on Orphan Well Association v Grant Thornton Ltd, 2019 SCC 5 [Blog post]. Retrieved from https://blogs.dal.ca/melaw/2019/02/22/reflections-on-orphan-well-association-v-grant-thornton-ltd-2019-scc-5/

Bodansky, D. (2017). The role of international court of justice in addressing climate change: Some preliminary reflections. *Arizona State law Journal, 49*, 689–712.

Brown, C., & Seck, S. (2013). Insurance law principles in an international context: Compensating losses caused by climate change. *Alberta Law Review, 50*(3), 541–576.

Burkett, M. (2009). Climate reparation. *Melbourne Journal of International Law, 10*(2), 509–542.

Burkett, M. (2015). Rehabilitation: A proposal for a climate compensation mechanism for small island states. *Santa Clara Journal of International Law, 13*(1), 81–124.

Byers, M., Franks, K., & Gage, A. (2017). The internationalization of climate damages litigation. *Washington Journal of Environmental Law & Policy, 7*(2), 264–319.

Centre for Climate Crime Analysis. Retrieved from http://www.climatecrimeanalysis.org/

Chambers, R., & Vastardis, A. Y. (2018). Overcoming the corporate veil challenge: Could investment law inspire the proposed business and human rights treaty. *International and Comparative Law Quarterly, 67*(2), 389–423.

Climate Action Tracker. Retrieved from https://climateactiontracker.org/

Climate Development and Knowledge Network (CDKN), International Centre for Climate Change and Development (ICCCD), Germanwatch, Munich Climate Insurance Initiative (MCII), & United Nations University – Institute for Human and Environment Security (UNU-HES). (2012). *Framing the loss and damage debate: A conservation starter*. Retrieved from https://germanwatch.org/sites/germanwatch.org/files/publication/6673.pdf

Clopton, Z. D. (2018). The global class action and its alternatives. *Theoretical Inquiries in Law, 19*, 125–150.

Collins, L., & McLeod-Kilmurray, H. (2014). *The Canadian law of toxic torts*. Toronto: Thomson Reuters Canada.

Desierto, D. (2018, February 14). Environmental damages, environmental reparations, and the right to a healthy environment: The ICJ compensation judgment in Cost Rica v Nicaragua and the IACtHR advisory opinion on marine protection for the greater Caribbean [Blog Post]. Retrieved from https://www.ejiltalk.org/environmental-damages-environmental-reparations-and-the-right-to-a-healthy-environment-the-icj-compensation-judgment-in-costa-rica-v-nicaragua-and-the-iacthr-advisory-opinion-on-marine-protection/

Diaz, S., Pascual, P., Stenseke, M., Martin-Lopez, B., Watson, R. T., Molnar, Z., ... Shirayama, Y. (2018). Assessing nature's contributions to people. *Science, 359*(6373), 270–272.

Durand, A., Hoffmeister, V., Weikmans, R., Gewirtzman, J., Natson, S., Huq, S., & Timmons Roberts, J. (2018). *Financing options for loss and damage: A review a roadmap* (discussion paper). Bonn, Germany: German Development Institute.

Ecocide Law Expert. Retrieved from https://pollyhiggins.com/

Expert Group on Climate Obligations of Enterprises. (2018). *Principles on climate obligations of enterprises*. Retrieved from https://climateprinciplesforenterprises.org

Fankhauser, S., Dietz, S., & Gradwell, P. (2014). *Noneconomic losses in the context of the UNFCCC work programme on loss and damage (policy paper)*. London, UK: Centre for Climate Change Economics and Policy – Grantham Research Institute on Climate Change and the Environment.

Gallmetzer, R. (2017). *Prosecute climate crimes*. Retrieved from UNEP http://web.unep.org/ourplanet/march-2017/articles/prosecute-climate-crimes

Ganguly, G., Setzer, J., & Heyvaert, V. (2018). If at first you don't succeed: Suing corporations for climate change. *Oxford Journal of Legal Studies, 38*(4), 841–868.

Gewirtzman, J., Natson, S., Richards, J. A., Hoffmeister, V., Durand, A., Weikmans, R., ... Roberts, J. T. (2018). Financing loss and damage: Reviewing options under the Warsaw international mechanism. *Climate Policy, 18*(8), 1076–1086.

Global Class Actions Exchange. Retrieved from http://globalclassactions.stanford.edu/

Government of Canada. (2018). Environmental damages fund. Retrieved from https://www.canada.ca/en/environment-climate-change/services/environmental-funding/programs/environmental-damages-fund.html

Greenpeace Philippines. (2015, September). Petition to the commission on human rights of the Philippines requesting for investigation of the responsibility of the carbon majors for human rights violations resulting from the impacts of climate change (Petition). Retrieved from https://www.greenpeace.org/seasia/ph/PageFiles/735291/CC%20HR%20Petition_public%20version.pdf

Heede, R. (2014). Tracing anthropogenic carbon dioxide and methane emissions to fossil fuel and cement producers 1854–2010. *Climatic Change, 122*, 229–241.

Humphreys, S. (2014). Climate justice: The claim of the past. *Journal of Human Rights and the Environment, 5*(Special Issue), 134–148.

International Bar Association. (2014). *Achieving justice and human rights in an era of climate disruption: Climate change justice and human rights task force report.* London, UK: Author.

IPCC. (2014). *Climate change 2014: Synthesis report. Contribution of working groups I, II and III to the Fifth assessment report of the intergovernmental panel on climate change.* Geneva, Switzerland: Author.

IPCC. (2018). *Global warming of 1.5°C – Special report: Summary for policymakers. Formally approved at the first joint session of working groups 1, II and III of the IPCC and accepted by the 48th session of the IPCC.* Incheon, Republic of Korea: Author.

Jam v International Finance Corp (S.C. 2019).

Jodoin, S., & Saito, Y. (2011). Crimes against future generations: Harnessing the potential of individual criminal accountability for global sustainability. *McGill International Journal of Sustainable Development Law and Policy, 7*(2), 115–155.

Juliana v United States of America 217 F.Supp.3d 1224 (D Or 2016), Aiken J.

Lees, E. (2017). Responsibility and liability for climate loss and damage after Paris. *Climate Policy, 17*(1), 59–70.

Leghari v Pakistan, W.P. No 25501/2015 No 1 (HC Green Bench, Pakistan, 2015). Retrieved from http://sys.lhc.gov.pk/greenBenchOrders/WP-Environment-25501-15-31-08-2015.pdf

Lord, R., Goldberg, S., Rajamani, L., & Brunnée, J. (Eds.). (2012). *Climate change liability: Transnational law and practice.* Cambridge, UK: Cambridge University Press.

Lyster, R. (2015). A fossil fuel-funded climate disaster response fund under the Warsaw international mechanism for loss and damage associated with climate change impacts. *Transnational Environmental Law, 4*, 125–151.

Marjanac, S., Patton, L., & Thornton, J. (2017). Acts of God, human influence and litigation. *Nature Geoscience, 10*, 616–619.

Mayer, B. (2014). Whose 'loss and damage'? Promoting the Agency of Beneficiary states. *Climate Law, 4*, 267–300.

McNamara, K. E., Bronen, R., Nishara Fernando, N., & Klepp, S. (2018). The complex decision-making of climate-induced relocation: Adaptation and loss and damage. *Climate Policy, 18*(1), 111–117.

Morrissey, J., & Oliver-Smith, A. (2013). *Perspectives on non-economic loss & damage: Understanding values at risk from climate change.* Retrieved from Loss and Damage in Vulnerable Countries Initiative http://www.eldis.org/document/A71918

Nishat, A., Mukherjee, N., Roberts, E., & Hasemann, A. (2013). *A range of approaches to address loss and damage from climate change impacts in Bangladesh.* Retrieved from http://asiapacificadapt.net/sites/default/files/resource/attach/a-range-of-approaches-to-address-loss-and-damage-from-climate-change-impacts-in-bangladesh.pdf

Ohdedar, B. (2016). Loss and damage from the impacts of climate change: A framework for implementation. *Nordic Journal of International Law, 85*(1), 1–36.

Olszynski, M., Mascher, S., & Doelle, M. (2018). From smokes to smokestacks: Lessons from tobacco for the future of climate change liability. *The Georgetown Environmental Law Review, 30*(1), 41.

Parker, H. R., Boyd, E., Cornforth, R. J., James, R., Otto, F. E. L., & Allen, M. (2017). Stakeholder perceptions of event attribution in the loss and damage debate. *Climate Policy, 17*(4), 533–550.

People of the State of New York v. Exxon Mobil Corporation, Docket number: 452044/2018, (N.Y. Sup. Ct. 2018). Retrieved from http://climatecasechart.com/case/people-v-exxon-mobil-corporation/

The Permanent Peoples' Tribunal. Retrieved from https://www.tribunalonfracking.org

Rayfuse, R. G., & Crawford, E. (2011). Climate change, sovereignty and statehood. In R. Rayfuse & S. Scott (Eds.), Edward Elgar. (2012). *Sydney Law School Research Paper No. 11/59.* Retrieved from Social Science Research Network Electronic library https://ssrn.com/abstract=1931466

Seck, S. L. (2000). Environmental harm in developing countries caused by subsidiaries of Canadian mining corporations: The interface of public and private international law. *Canadian Yearbook of International Law/Annuaire Canadien de Droit International, 37*, 139–221.

Seck, S. L. (2017). Revisiting transnational corporations and extractive industries: Climate justice, feminism, and state sovereignty. *Transnational Law & Contemporary Problems, 26*(2), 383–413. (Symposium: International Environmental Law, Environmental Justice, and the Global South).

Serdeczny, O. M., Bauer, S., & Huq, S. (2018). Non-economic losses from climate change: Opportunities for policy-oriented research. *Climate and Development, 10*(2), 97–101.

Siegele, L. (2017). Loss and damage (Article 8). In D. Klein, M. P. Carazo, M. Doelle, J. Bulmer, & A. Higham (Eds.), *The Paris agreement on climate change: Analysis and commentary* (pp. 224–238). Oxford, UK: Oxford University Press.

Stabinsky, D., & Hoffmaister, J. P. (2012). *Loss and damage: Defining slow onset events* (Briefing Paper 3). Retrieved from http://unfccc.int/files/adaptation/application/pdf/tp7_v03_advance_uneditted_version.pdf

Stephens, T. (2009). *International courts and environmental protection.* Cambridge, UK: Cambridge University Press.

Stone, C. D. (2010). *Should trees have standing? Law, morality, and the environment* (3rd ed.). Oxford, UK: Oxford University Press.

Strauss, A. (2009). Climate change litigation: Opening the door to the international court of justice. In W. C. G. Burns & H. M. Osofsky (Eds.), *Adjudicating climate change: State, national, and international approaches* (pp. 334–356). Cambridge, UK: Cambridge University Press.

Trudel, Johnston & Lespérance. Ongoing class actions: *Environment JEUnesse v Attorney general of Canada.* Retrieved from http://tjl.quebec/en/class-action/climate-change/

Tsosie, R. (2007). Indigenous people and environmental justice: The impact of climate change. *University of Colorado Law Review, 78*, 1625.

UNEP. (2016). *Loss and damage: The role of ecosystem services.* Nairobi, Kenya: UNEP. Retrieved from https://uneplive.unep.org/media/docs/assessments/loss_and_damage.pdf

UNFCCC. (2014). *Report of the Warsaw international mechanism for loss and damage associated with climate change impacts* (UNFCCC document FCCC/SB/2014/4). Paris, France. Retrieved from http://unfccc.int/resource/docs/2014/sb/eng/04.pdf

United Nations. (2001). *Draft articles on responsibility of states for internationally wrongful acts with commentaries* (Test adopted by ILC at its 53rd session A/56/10). Retrieved from http://legal.un.org/ilc/texts/instruments/english/commentaries/9_6_2001.pdf

United Nations. (2006). *Draft principles on the allocation of loss in the case of transboundary harm arising out of hazardous activities, with commentaries* (Text adopted by ILC at its 58th session A/61/10). Retrieved from http://legal.un.org/ilc/texts/instruments/english/commentaries/9_10_2006.pdf

United Nations. (2007). *Declaration on the rights of indigenous peoples.* UN General Assembly Resolution. Retrieved from https://undocs.org/A/RES/61/295

United Nations. (2014). Decision 2/ CP.19, *Warsaw international mechanism for loss and damage associated with climate change impacts,* FCCC/ CP/ 2013/ 10/ Add.1.

United Nations. (2018). *Decision -/CMA.1: Matters relating to the implementation of the Paris Agreement* (advance unedited version). Retrieved from https://unfccc.int/sites/default/files/resource/cp24_auv_3cma1_final.pdf

Urgenda Foundation v The Netherlands (Ministry of Infrastructure and Environment) C/09/456689/HA ZA 13-1396 (District Court, The Hague, 2015), english translation. Retrieved from https://uitspraken.rechtspraak.nl/inziendocument?id=ECLI:NL:RBDHA:2015:7196

Van Den Homberg, M., & McQuistan, C. (2019). Technology for climate justice: A reporting framework for loss and damage as part of key global agreements. In R. Mechler, L. Bouwer, T. Schinko, S. Surminski, & J. Linnerooth-Bayer (Eds.), *Loss and damage from climate change: Concepts, methods and policy options?* (pp. 513–545). Cham: Springer, version. https://doi.org/10.1007/978-3-319-72026-5_22

Vedanta Resources PLC v Lungowe [2019] UKSC 20.

Verheyen, R. (2012). *Loss and damage: Tackling loss & damage – A new role for the climate regime?* Retrieved from Loss and Damage in Vulnerable Countries Initiative http://www.lossanddamage.net/download/6877.pdf

Voigt, C. (2016). The potential role of the international court of justice in climate change-related claims. In M. Faure (Ed.), *Elgar Encyclopedia of environmental law.* Northampton: Edward Elgar Publishing.

Wanless, W. C. (2009). Corporate liability for international crimes under Canada's crimes against humanity and war crimes act. *Journal of International Criminal Justice, 7,* 201–221.

Watt-Cloutier, S. (2016). *The right to be cold: One woman's story of protecting her culture, the arctic, and the Whole Planet.* Toronto: Penguin Random House Canada.

Wewerinke-Singh, M. (2018). State responsibility for human rights violations associated with climate change. In S. Duyck, S. Jodoin, & A. Johl (Eds.), *Routledge Handbook of human rights and climate governance.* London & New York: Routledge.

World Health Organization (WHO). (2018). *COP 24 special report: Health and climate change.* Katowice, Poland: United Nations. Retrieved from https://www.who.int/globalchange/publications/COP24-report-health-climate-change/en/

Wrathall, D., Oliver-Smith, A., Fekete, A., & Sakdapolrak, P. (2015). Problematising loss and damage. *International Journal of Global Warming, 8*(2), 274–294.

Yaiguaje v. Chevron Corporation, 2018 ONCA 472, 423 DLR (4th) 687.

Zommers, Z., Wrathwall, D., & Can der Geest, K. (2014). *Loss and damage to ecosystem services UNU-EHS working paper series. No. 12.* Bonn, Germany: United Nations Institute for Environment and Human Security.

ẟ OPEN ACCESS

Between negotiations and litigation: Vanuatu's perspective on loss and damage from climate change

Margaretha Wewerinke-Singh ⓘ and Diana Hinge Salili ⓘ

ABSTRACT

This contribution explores how climate-vulnerable states can effectively use the law to force action in order to address loss and damage from climate change, taking the Pacific Island state of Vanuatu as an example. Vanuatu made headlines when its Minister of Foreign Affairs, International Cooperation and External Trade, the Hon. Ralph Regenvanu, announced his government's intention to explore legal action as a tool to address climate loss and damage suffered in Vanuatu. Our contribution places this announcement in the context of Vanuatu's own experience with climate loss and damage, and the state's ongoing efforts to secure compensation for loss and damage through the multilateral climate change regime. We then discuss the possibilities for legal action to seek redress for climate loss and damage, focusing on two types of action highlighted in Minister Regenvanu's statement: action against states under international law, and action against fossil fuel companies under domestic law. After concluding that the issue of compensation for climate loss and damage is best addressed at the multilateral level, we offer proposals on how the two processes of litigation and negotiation could interact with each other and inspire more far-reaching action to address loss and damage from climate change.

Key policy insights

- The review of the Warsaw International Mechanism for Loss and Damage offers an opportunity to start putting in place a facility for loss and damage finance under the auspices of the United Nations Framework Convention on Climate Change (UNFCCC).
- A climate damages tax (CDT) on fossil fuel companies seems a particularly promising option for mobilizing loss and damage finance. Such a CDT could be one revenue stream for a relevant loss and damage facility.
- Legal action – including cases against foreign states or fossil fuel companies – could bolster the position of climate-vulnerable states in multilateral negotiations on loss and damage finance.

Introduction

This contribution explores how climate-vulnerable states such as the Pacific Island state of Vanuatu can effectively use the law to forge action to address loss and damage. Vanuatu made headlines when its Minister of Foreign Affairs, International Cooperation and External Trade, the Hon. Ralph Regenvanu, announced the Government of Vanuatu's intention to explore legal action against the fossil fuel industry, and the states that sponsor it, for climate change-related loss and damage the state had suffered (Republic of Vanuatu, 2018c). This announcement must be seen in the context of the severe loss and damage already suffered in Vanuatu and across the globe due to climate change, accompanied by the failure of the multilateral climate change

This is an Open Access article distributed under the terms of the Creative Commons Attribution License (http://creativecommons.org/licenses/by/4.0/), which permits unrestricted use, distribution, and reproduction in any medium, provided the original work is properly cited.

regime to provide compensation for such loss and damage. While compensation has been an extremely controversial issue at the international climate change negotiations (Pekkarinen, Toussaint, & Van Asselt, 2019; Roberts & Huq, 2015), it is based on climate justice imperatives that are widely recognized in the literature (Burkett, 2014; Farber, 2008; Tschakert, Ellis, Anderson, Kelly, & Obeng, 2019; Vanhala & Hestbaek, 2016; Wallimann-Helmer, 2015; Warner & Zakieldeen, 2011). The call for compensation for loss and damage is also supported by well-established rules and principles of international law, including the right to reparations for injury resulting from violations of international law (Hafner-Burton, Victor, & Lupu, 2012; Wewerinke-Singh, 2019).

In line with the Intergovernmental Panel on Climate Change (IPCC)'s Special Report on Global Warming of 1.5°C, we understand climate loss and damage broadly as harms from (observed) impacts and (projected) risks associated with future climate change (IPCC, 2018, Glossary, p. 553). However, unlike the IPCC (2018), but in line with the Paris Agreement (2015) and other relevant documents that form part of the international climate change regime under the United Nations Framework Convention on Climate Change (UNFCCC), we do not distinguish the political debate on Loss and Damage (capitalized letters) from physical losses and damages (lowercase letters) at the definitional level, and use the term 'loss and damage' to refer to both. We conceptualize compensation as 'something, typically money, awarded to someone in recognition of loss, suffering or injury' (Oxford Living Dictionary, 2018). We refer to Vanuatu as a useful illustrative case that has lessons for other climate-vulnerable states seeking to catalyze meaningful global action on loss and damage, including compensation.

Vanuatu's submission to the Executive Committee of the Warsaw International Mechanism (WIM) for Loss and Damage of the UNFCCC on loss and damage finance provides an indication of the range of activities that could be covered by compensation for loss and damage: costs of relocation due to sea-level rise for coastal communities; costs for climate resilient reconstruction after extreme weather events; social and gender protection measures; livelihood safety net programmes for the most vulnerable; livelihood transformation programmes; pro-poor micro insurance, crop insurance and/or insurance premium subsidies at various levels; national and local level emergency finance reserves or contingency funds; contingency planning and comprehensive risk management particularly at the local level; capacity and institution building at all levels; and technology cooperation and transfer, such as loss and damage assessment tools (Republic of Vanuatu, 2018a). This wide range of activities illustrates that compensation for loss and damage would need to address both quantifiable damages and intangible, non-economic losses (Tschakert et al., 2019).

Vanuatu's experience with various types of loss and damage is discussed in the first part of the article, which serves to illustrate the urgency of the issue and the necessity of international cooperation to address it. Vanuatu's attempts to secure compensation for loss and damage through multilateral negotiations are discussed in the second part, while the third part highlights the problems of moving the negotiations forward on this issue. The fourth part discusses the possibilities for legal action to seek redress for loss and damage, focusing on the two types of action highlighted in Minister Regenvanu's statement: action against states under international law, and action against fossil fuel companies under domestic law. The conclusion then brings the different lines of analysis together and leads us to make proposals as to how the two processes of litigation and negotiation could interact with each other and inspire more ambitious action to address loss and damage.

Vanuatu's experience with loss and damage

Vanuatu is composed of 82 islands that are mostly of volcanic origin, 65 of which are inhabited by people.[1] Its population of around 272,000 depends mainly on agriculture and fishing for subsistence (Vanuatu National Statistics Office (VNSO), 2017). The UN categorizes Vanuatu as a Least Developed Country (LDC), largely due to its being one of the most geo-physically vulnerable countries in the world (Garschagen et al., 2015; UN Department of Economic and Social Affairs, 2010). Vanuatu is located in the 'Pacific Rim of Fire', experiencing moderate earthquakes almost every other day, and regular eruptions from more than a dozen active volcanoes (Vanuatu Metrology and Geo-Hazards Department [Vanuatu Meteo], 2017). The archipelago also falls within the South Pacific tropical cyclone basin and is subject to frequent cyclones, tsunamis, storm surges, drought, flooding of both coasts and rivers, and landslides (Pacific Catastrophe Risk Assessment and Financing Initiative

[PCRAFI], 2011). The state's geographical vulnerability is exacerbated by low levels of economic development and institutional capacity (Global Facility for Disaster Reduction and Recovery [GFDRR], 2009). From 2011 to 2015, Vanuatu was ranked as the world's most vulnerable nation to natural disasters on the World Risk Index (Garschagen et al., 2015). Under the UNFCCC regime, Vanuatu qualifies as 'particularly vulnerable to the adverse effects of climate change' due to its status as a 'small island country' (UNFCCC, preamble).

Vanuatu has seen an increase in the frequency and intensity of extreme weather events due to climate change (GFDRR, 2009; Government of Vanuatu, 2018a). It is expected that natural disasters and extreme weather events will intensify even further as temperatures continue to rise (GFDRR, 2009). Climate modelling for Vanuatu has predicted increased average temperatures and an increase in the frequency of extreme temperatures; both greater precipitation and a higher number of dry days during the rainy and dry season respectively; sea level rise; and coastal erosion (Vanuatu Meteo, 2017). The complex and damaging interaction between climate change impacts contributed to the destructive capacity of category five Cyclone Pam, which struck Vanuatu in March 2015. Cyclone Pam was the largest ever recorded in the South Pacific, until Cyclone Winston which hit Fiji the following year (Sopko & Falvey, 2015). At least 16 people lost their lives as a result of Cyclone Pam (Australian Broadcasting Cooperation, 2015), while approximately 65,000 people (roughly one quarter of the population) were displaced by the disaster. The Government of Vanuatu estimates that some 17,000 buildings were damaged or destroyed (Esler, 2015). The destruction and subsequent economic losses totalled approximately US$450 million, or roughly 64% of GDP (Vanuatu Meteo, 2017). The impact of the disaster was amplified by a subsequent drought associated with the unusually strong 2015–2016 El Niño event, the intensity of which has also been linked to climate change (Cho, 2016).

The extended impact of the disaster demonstrates how climate events undermine economic and social development and can lead to poverty traps (Heltberg, Siegel, & Jorgensen, 2009). As New Zealand's Prime Minister Jacinda Ardern witnessed personally when visiting Vanuatu in March 2018, some Ni-Vanuatu schoolchildren still attend class in tents because the school buildings damaged by the storm were neither replaced nor repaired (Ardern, 2018). Some children have dropped out of school altogether. The extent of the disruption to Vanuatu's economy and society has caused its graduation from LDC status, scheduled for December 2017, to be postponed until December 2020 (UN Office of the High Representative for the Least Developed Countries, Landlocked Developing Countries and Small Island Developing States, 2018). Minister Regenvanu, lamented in his high-level statement to the 24th Conference of the Parties (COP24) to the UNFCCC in Katowice, Poland, in 2018 that the adverse effects of climate change leave Vanuatu in a permanent state of emergency (Government of Vanuatu, 2018b).

Vanuatu will almost certainly suffer additional large-scale and devastating losses and damages as temperatures continue to rise. According to PCRAFI (2011), Vanuatu is expected to incur, on average over the long term, losses of US$48 million per year due to tropical cyclones and earthquakes, which is equivalent to 6.6% of Vanuatu's GDP. PCRAFI (2011) further estimates that Vanuatu has a 50% chance of experiencing a loss exceeding US$330 million and casualties larger than 725 people from a single event in the next 50 years. These numbers do not yet account for loss and damage resulting from slow-onset climate disasters such as drought or coastal erosion, or complex interactions between different types of impacts. More research is needed to better understand the severity and nature of the risks of future loss and damage for Vanuatu under different climate scenarios. However, it is clear from Vanuatu's experience with Cyclone Pam that the costs of future loss and damage will far exceed the nation's coping capacities. This is even more so if the costs of preventing and addressing non-economic loss and damage – such as the loss of traditional knowledge or distinct ways of life – are also taken into account. Vanuatu has therefore intensified its efforts to acquire loss and damage finance from international sources through the multilateral loss and damage regime.

Vanuatu's role in the emerging multilateral loss and damage regime

Vanuatu's efforts to secure compensation for the damaging consequences of climate change started when the UNFCCC was negotiated. At the Intergovernmental Negotiating Committee, the body established by the UN

General Assembly (UNGA) to negotiate the UNFCCC, the Alliance of Small Island States (AOSIS) chaired by Vanuatu's ambassador Robert Van Lierop put forward a proposal in 1991 to establish an international fund and insurance pool based on the polluter pays principle. The proposal explained how '[t]he resources of the insurance pool should be used to compensate the most vulnerable small island and low-lying coastal developing countries for loss and damage resulting from sea level rise' (AOSIS, 1991; Linnerooth-Bayer, Mace, & Verheyen, 2003). Whilst the proposal was rejected by developed states, AOSIS did secure a reference to insurance in Article 4.8 of the Convention (UNFCCC, 1992). Similar language, including 'insurance', was included in Article 3.14 of the Kyoto Protocol (Kyoto Protocol, 1997).

It would take another fifteen years until the first reference to 'loss and damage' appeared in a decision of the COP, entitled the Bali Action Plan (UNFCCC, 2007, para. 1(c)(iii)). The term has since appeared in numerous COP decisions, and in the name and mandate of the Warsaw International Mechanism on Loss and Damage (WIM) established in 2013. The WIM is mandated to undertake, inter alia, the function of '[e]nhancing action and support, including finance, technology and capacity-building, to address loss and damage associated with the adverse effects of climate change' (UNFCCC, 2013, para. 5(c)). As Vanhala and Hestbaek explain, it was the ambiguity of the 'loss and damage' framing that unlocked progress in the negotiations on the issue (Vanhala & Hestbaek, 2016). On the one hand, the lack of a reference to either 'liability' or 'compensation' accommodated developed states' concerns about their potential liability for the consequences of climate change (Huq, Roberts, & Fenton, 2013; Roberts & Huq, 2015). On the other hand, developing states saw a breakthrough in the establishment of a process to address loss and damage from climate change (Roberts & Huq, 2015). A specific victory for Vanuatu and other developing states was the inclusion of a reference to 'recovery and rehabilitation' as one method to remedy loss in the initial two-year workplan of the WIM Executive Committee adopted at COP20 in Lima in 2014 (UNFCCC, 2014, p. 10), and in its five-year rolling workplan adopted at COP23 in Bonn in 2017 (UNFCCC, 2017a).

The 2015 Paris Agreement sharpened the edges of the emerging multilateral loss and damage regime. Emboldened by demands from Pacific civil society, Vanuatu, alongside other developing states, had advocated for the inclusion of a standalone article on loss and damage in the Paris Agreement that would also 'anchor' the WIM, or a new mechanism with a broader mandate, into the climate change regime (SPREP, 2015). This was achieved through what is now Article 8 of the Paris Agreement, which provides the WIM with a durable legal basis while allowing for its enhancing and strengthening (Paris Agreement, 2015). Further, paragraph 3 of Article 8 explicitly directs parties 'to enhance understanding, action and support' to address loss and damage. However, the latter provision lacks an explicit link to the financial mechanism of the Convention; this omission creates uncertainty about whether loss and damage is eligible for funding from the UNFCCC's main climate fund, the Green Climate Fund (GCF) (Siegele, 2018). Developed states' resistance to the inclusion of a provision on loss and damage in the Paris Agreement, combined with developing states' eagerness to find a workable compromise (Hoffmaister, Talakai, Damptey, & Barbosa, 2014) resulted in another important limitation of the emerging regime. This limitation is reflected in paragraph 52 of the decision accompanying the Paris Agreement, which stipulates that 'Article 8 of the Agreement does not involve or provide a basis for any liability and compensation' (UNFCCC, 2015a). As this paragraph is important to understanding the relationship between negotiations and litigation, and Vanuatu played an important role in negotiating it, it is worth unpacking its origins and implications.

When considering the origins of paragraph 52, a key point to note is that the scope of the provision is narrower than what was initially proposed by its main architect, the US. The US had wanted to exclude liability and compensation from the scope of the international climate change regime altogether (Calliari, 2018; Pekkarinen et al., 2019). Text reflecting this desire appeared in a 'Draft Paris Outcome' prepared by the French COP Presidency towards the end of the negotiations in Paris, which suggested a treaty provision on loss and damage of which the third and final paragraph would have read as follows:

> Parties shall enhance action and support, on a cooperative and facilitative basis, for addressing loss and damage associated with the adverse effects of climate change, and in a manner that does not involve or provide a basis for liability or compensation nor prejudice existing rights under international law. (UNFCCC, 2015b)

The inclusion of this clause within the Paris Agreement itself would have meant that liability and compensation could only have been introduced (or re-introduced, depending on one's perspective) into the international climate change regime through a formal treaty amendment, which would be an extremely cumbersome process with negligible chances of a successful outcome. The provision in the Presidency's Draft Outcome therefore crossed Vanuatu's red lines as well as the red lines of a number of other states, including other Small Island Developing States (SIDS) in AOSIS and members of the African Group. To address this issue, Vanuatu's head of delegation arranged an emergency meeting with the Presidency during the final hours of the conference to indicate that Vanuatu would be unable to join consensus on the Paris Agreement if the proposed clause were included in the treaty text. The clause was abandoned. The paragraph that was eventually included in decision 1/CP.21 was an acceptable (though still undesirable) compromise for Vanuatu (Mace & Verheyen, 2016; Wewerinke-Singh & Doebbler, 2016).

The implications of the exclusion of liability clause in decision 1/CP.21 are not as sweeping as they may at first appear. A first point to underscore is that, unlike the exclusion of liability clause contained in decision 1/CP.21, the broadly-drafted language of Article 8 of the Paris Agreement cannot be altered without a formal treaty amendment. Secondly, despite the clause contained in decision 1/CP.21, the mandate of the WIM is still 'broad enough to encompass many of the concerns addressed by what has been termed "compensation"' (Mace & Verheyen, 2016, p. 210). Indeed, the WIM Executive Committee's five-year rolling workplan adopted in 2017 explicitly states that it will implement the WIM's function of '[e]nhancing action and support, including finance, technology and capacity-building, to address loss and damage associated with the adverse effects of climate change' (UNFCCC, 2017a). The five-year rolling workplan also contains a strategic workstream focused on enhancing cooperation and facilitation in relation to this function (UNFCCC, 2017a). Moreover, as Lees (2017) explains, the exclusion of liability clause leaves open the possibility of assigning legal responsibility for loss and damage under the international climate change regime. Herein lies much of the untapped potential of the WIM: the development of a responsibility allocation mechanism would enable states 'to utilize this mechanism to transfer their own responsibilities onto private actors' (Lees, 2017, p. 68).

A final point to mention is that upon signing the Paris Agreement, Vanuatu – along with a number of other climate-vulnerable states – made a formal declaration stating that 'ratification of the Paris Agreement shall in no way constitute a renunciation of any rights under any other laws, including international law' (Republic of Vanuatu, 2018b). This declaration reaffirms the entitlement of climate vulnerable states to pursue compensation for climate loss and damage through legal action outside the UNFCCC process (Pekkarinen et al., 2019). It is worth recalling that several SIDS had already made a similar declaration under the UNFCCC (1992) and the Kyoto Protocol (1997). The potential importance of legal action to address loss and damage has increased since the adoption of the Paris Agreement, as some developed states are deliberately blocking the operationalization of the regime in relation to loss and damage finance (Pekkarinen et al., 2019). This situation and its implications are discussed in the following section.

The intensifying battle over loss and damage finance

While negotiations on loss and damage give the appearance of some progress at nearly every COP (and some intersessionals), the overall picture since Paris is one of stagnation. At COP23 in 2017, Vanuatu and most other developing states made a strong call for new and additional finance for loss and damage, and suggested the inclusion of a permanent item on loss and damage in the agenda of the subsidiary bodies. Developed states opposed both proposals, with the US arguing with renewed vigour against the operationalization of the loss and damage regime (Benjamin, Thomas, & Haynes, 2018). The parties eventually reached a compromise in calling for a one-off 'expert dialogue' to explore how loss and damage finance may be mobilized and secured (UNFCCC, 2017b).

In the run-up to the expert dialogue, branded the 'Suva Expert Dialogue on Loss and Damage' by the Fijian COP23 Presidency, Vanuatu's Minister Regenvanu started advocating in public for a 'climate damages tax' (CDT) on fossil fuel companies. Minister Regenvanu first made a call for a CDT at a meeting of Commonwealth Heads of Government in London in April 2018 (Darby, 2018), followed by the publication of several co-authored opinion pieces in popular online outlets (Isaac, Jumeau, Mahmud, & Regenvanu, 2018; Regenvanu & Persaud, 2018). According to Minister Regenvanu and Persaud (2018), a CDT is based on the rationale that

> [The fossil fuel industry] has spent decades fueling climate denial while making profits. In 2017 alone, the top six oil companies made $134 billion in profit. … We will only stop climate change by making those who contribute to it pay for it. … We need to end the mismatch between those who gain and those who lose.

NGOs supporting this rationale have defined CDT as 'a charge on the extraction of each tonne of coal, barrel of oil, or cubic litre of gas, calculated at a consistent rate globally based on how much climate pollution (CO2e) is embedded within the fossil fuels' (Richards, Hillman, & Boughey, 2018, p. 3). The proposed CDT would also see the royalties paid to states by fossil fuel companies channelled to a loss and damage facility managed by the GCF (Richards et al., 2018). This would be beneficial in anchoring loss and damage finance in the overarching framework of the UNFCCC's financial mechanism to minimize the complexity of accessing climate finance (Republic of Vanuatu 2018a).

A crucial aspect of the proposal is that developed states are not asked to provide finance from public sources to compensate climate-vulnerable states for loss and damage, but instead to adopt legislation that would make it mandatory for private actors from the fossil fuel industry to provide such compensation (Frumhoff, Heede, & Oreskes, 2015). However, when the proposal was put forward by developing states and observers at the Suva Expert Dialogue in May 2018 (UNFCCC, 2018a), it was opposed by developed states (Singh, 2018). At COP24, parties merely welcomed the report of the Suva Expert Dialogue and encouraged the WIM's Executive Committee '[t]o seek ways to continue enhancing its responsiveness, effectiveness and performance in implementing activities in its five-year rolling workplan, particularly those under [the workstream on action and support]' (UNFCCC, 2018b). Moreover, parties failed to expressly acknowledge Article 8 alongside mitigation and adaptation in the 'rulebook' of the Paris Agreement. As Pekkarinen et al. (2019, p. 36) note, this was 'a missed opportunity' and 'further contributes to the confusion surrounding the differences between adaptation and loss and damage'.

At the time of writing, an important contribution to the loss and damage finance discussion is expected in the form of a technical paper which the secretariat is mandated to prepare as an input to the review of the WIM at COP25 in 2019. This paper is to identify the sources of financial support for addressing loss and damage as well as the modalities for accessing such support (UNFCCC, 2018b). The results of the Suva Expert Dialogue, including the proposal for a CDT, will be used to inform this paper. Thus, the paper has the potential to inform substantive negotiations about the proposal. At the same time, the review of the WIM provides an opportunity to start a process to provide this body with the envisaged financial arm (Pekkarinen et al., 2019; Richards et al., 2018). However, the political dynamics around loss and damage finance warrant skepticism about the prospects for progressing these proposals through the climate negotiations (Benjamin et al., 2018). Against this backdrop, it is urgent to understand how legal action could help to change the dynamics and potentially contribute to a comprehensive multilateral agreement on loss and damage with enforceable provisions on climate finance.

Legal action to pursue compensation for climate loss and damage

At the Climate Vulnerable Summit in November 2018, Minister Regenvanu announced his government's intention to explore legal action against the corporations and governments profiting most from products causing climate change, stating that it was

> exploring all avenues to utilize the judicial system in various jurisdictions – including under international law – to shift the costs of climate protection back onto the fossil fuel companies, the financial institutions and the governments that actively and knowingly created this existential threat to Vanuatu. (Republic of Vanuatu, 2018c)

Legal action under international law is premised on substantive obligations relevant to climate change derived from the law of nations. This form of legal action has been contemplated by climate-vulnerable states for nearly two decades. In 2002, Tuvalu's then Prime Minister Koloa Talake announced that Tuvalu was considering bringing a contentious case against Australia and the US for their alleged failure to address global warming. Subsequently, the then Tuvaluan Minister of Finance, Bikenibeu Paeniu tried to build a coalition of SIDS from the

Pacific, the Indian Ocean and the Caribbean to join Tuvalu in the planned lawsuit (Jacobs, 2005; Reuters, 2002). However, the initiative was dropped following a change of government in Tuvalu. Nearly ten years later, in 2011, the then President of Palau, Johnson Toribiong, called on the UNGA to

> seek, on an urgent basis … an advisory opinion from the International Court of Justice on the responsibilities of states under international law to ensure that activities carried out under their jurisdiction or control that emit greenhouse gases do not damage other states. (Hurley, 2011)

This initiative was dropped as well, reportedly due to threats of reprisal by the US with which Palau has close ties under a Compact of Free Association (Brown, 2013; Burkett, 2013).

In addition to likely opposition from powerful states, there are, not surprisingly, several legal obstacles to successful climate litigation before an international body such as the International Court of Justice (ICJ). These include finding a suitable forum; identifying relevant substantive obligations and various challenges relating to attribution, causation and evidence. These obstacles are not inherently insurmountable (Rajamani, 2015, p. 19; Verheyen, 2005; Wewerinke-Singh, 2019) and legal action could be designed to circumvent at least some of them. Vanuatu could, for example, accept the compulsory *ipso facto* jurisdiction of the ICJ under Article 36, paragraph 2 of the ICJ Statute and then bring a test case before the ICJ against one or more states that have done the same, which would include Australia but not the US. The case could focus on states' prevention obligations under the *lex specialis* of the UNFCCC, human rights law or customary international law. An alternative approach could be to revive Palau's earlier campaign for a UNGA resolution requesting the ICJ for an advisory opinion on climate change, or to seek an advisory opinion on loss and damage relating to marine ecosystems from the International Tribunal for the Law of the Sea (ITLOS) (Bodanksy, 2017; Sands, 2016).

The second type of legal action contemplated in Minister Regenvanu's statement focuses on the contribution of the world's largest investor-owned fossil fuel producers to loss and damage. Frumhoff et al. (2015) identify four reasons why these so-called 'carbon majors' carry significant responsibility for climate change. Firstly, their products are responsible for a large share of the anthropogenic greenhouse gas emissions accumulated in the global atmosphere. Secondly, they continued to produce these products despite recognition of the danger by scientists and policy-makers. Thirdly, they have systematically undermined political action on climate change mitigation, including through the promotion of misinformation. Finally, they continue to work towards the expanded production and use of fossil fuels for decades to come. This multi-faceted responsibility has inspired a growing number of legal cases against the carbon majors brought in domestic courts around the world, notably including suits brought by cities and local governments on behalf of their citizens (Ganguly, Setzer, & Heyvaert, 2018). Plaintiffs in these cases are invoking a wide range of legal theories, ranging from public and private nuisance through to trespass (Burger & Wentz, 2018). While the outcomes of these cases are difficult to predict, the precedent of tobacco lawsuits illustrates that it is possible, at least in principle, for public authorities to recover billions of dollars from industry on behalf of their citizens (Ieyoub and Eisenberg, 2000).

In Vanuatu, a case against carbon majors could be premised on domestic tort law, with jurisdiction flowing from the fact that the claim involves harm that occurred (and is ongoing) within Vanuatu's borders (Gage & Wewerinke, 2015). A similar model has been used in a petition to the Commission on Human Rights of the Philippines, which instigated a National Inquiry on Climate Change to investigate the responsibility of 47 investor-owned carbon majors for human rights violations resulting from climate change. At the time of writing, the Commission was in the process of completing its recommendations from the inquiry (R. Cadiz, personal communication, March 4, 2019). While the Commission does not have the power to produce binding orders, the process itself has already contributed to enhanced accountability of the actors most responsible for climate change and recognition of its human rights consequences (Savaresi, Hartmann, & Cismas, 2019).

Vanuatu does not, as yet, have its own national human rights institution with a mandate to conduct a similar inquiry. However, it does have an independent and highly regarded judicial system that could potentially deal with a claim for climate damages against carbon majors. The chances of obtaining a favourable judgment or even a settlement in such a case are regarded as highly uncertain, as climate litigation against fossil fuel companies comes with its own significant hurdles. These include the challenge of establishing causation between the defendants' conduct on the one hand and the specific damages asserted by Vanuatu on the other (Burger &

Wentz, 2018; Kysar, 2011), which is particularly difficult in connection with non-economic loss and damage (Tschakert et al., 2019). Nonetheless, this type of litigation could potentially offer Vanuatu a real chance to secure significant financial compensation for climate damages from those responsible for causing it.

Conclusion

Against the backdrop of increasingly severe loss and damage suffered in climate-vulnerable states on the one hand, and stagnating multilateral negotiations on the other, the potential of legal action focused on loss and damage is gaining renewed importance. Legal action is not an ideal strategy to address loss and damage: each legal initiative remains piecemeal, and as such does not substitute for a comprehensive and enforceable multilateral agreement on the issue. Moreover, each form of legal action comes with potentially significant risks and costs, such as the risk of creating an adverse precedent or facing reprisals from powerful corporate or government defendants. Nonetheless, Vanuatu and other climate-vulnerable states may consider that the potential benefits of legal action outweigh the risks, especially if those risks are mitigated through a carefully crafted case strategy and coalition-building.

As Peel and Osofsky (2015) point out, part of the added value of climate litigation consists in its capacity to produce public debate over difficult and controversial questions that are otherwise too easily swept under the carpet. To list but a few pressing questions that have been left unaddressed by the multilateral climate change regime: Which actors are primarily responsible for loss and damage from climate change, and what role should they play in addressing it? What is the standard for determining that specific loss and damage is attributable to anthropogenic climate forcings? How can uncertainty be addressed in this regard? What kinds of remedies could best restore the rights of those who suffer loss and damage from climate change? How should our laws and societies address climate-related irreversible loss? Given the global governance implications of these questions, they are best addressed at the multilateral level. However, litigation would allow Vanuatu and other climate-vulnerable states to expose these issues more openly, fully and forcefully. At the same time, it could provide individuals and communities with first-hand experience of loss and damage with an opportunity 'to be heard, accuse and explain' (Duffy, 2018, p. 51). Linking peoples' lived experiences of loss and damage with the question of responsibility through a legal case would almost certainly inspire a more meaningful international debate on loss and damage, and may even result in more ambitious action to prevent and address it.

State-based international litigation appears to have the greatest potential to influence the multilateral negotiations directly. Contentious cases can provide heightened exposure to the conduct of some (though not all) recalcitrant states, with the potential to result in a binding judgment that could include an order to make full reparations for injury suffered as a result of wrongful conduct (Wewerinke-Singh, 2019). Advisory opinions, while not binding, have the potential at least to clarify the rights and obligations of all states in connection with loss and damage (Rajamani, 2015). Increased clarity about states' rights and obligations could in turn bolster the position of climate-vulnerable states in international climate negotiations (Schwarte & Byrne, 2011). The influence of an international judgment or advisory opinion on climate change could extend to other areas of international law, including trade negotiations under the auspices of the World Trade Organization and international investment arbitration. Moreover, it could assist citizens, local governments and organizations around the world in holding recalcitrant states to account before regional and domestic courts (Bodanksy, 2017; Schwarte & Byrne, 2011) which could in turn affect states' positions in the multilateral negotiations. Coalition-building is critical to avoid reprisals and, in case of an advisory opinion, to secure a majority of votes in the UNGA. In particular, those states which, like Vanuatu, made declarations reserving their rights under international law upon signing or ratifying the Paris Agreement are potential allies in crafting an effective strategy to address loss and damage through legal action.

Litigation against fossil fuel companies is another strategy that climate-vulnerable states could pursue in the quest for climate justice. Indeed, the filing of the first case against carbon majors by a sovereign nation would powerfully underscore the risks of mounting liabilities to which these companies have exposed themselves through their actions and omissions that contribute to climate change. Increased exposure to litigation for losses from climate change already appears to have motivated companies to reduce their emissions, and institutional investors to divest from fossil fuel companies (United Kingdom Sustainable Investment and Finance Association and Climate Change Collaboration, 2018). Moreover, as Hunter (2009) points out, the exposure to litigation could ultimately inspire industry to promote a liability regime under the UNFCCC that would mitigate the risk of liability through a political compromise. In this way, litigation against carbon majors could indirectly address the lack of political will to address loss and damage on the part of the states where those corporations are headquartered.

In sum, this article has made clear that the question facing climate-vulnerable states is not whether, but how, to use the law to force meaningful action to address climate change-related loss and damage. Answering this question will require further deliberations amongst climate-vulnerable states that take account of their individual circumstances as well as their collective objectives in multilateral negotiations. Ultimately, a combination of legal initiatives and diplomacy may offer the greatest chances of catalyzing transformative change at the global level and obtaining much-needed reparations for actual climate harm.

Note

1. This section draws in part on Wewerinke-Singh and Van Geelen (2019).

Acknowledgements

The authors are grateful to Ambassador John Licht, Jesse Benjamin, Robson Tigona, Daniel Galpern, Jeff Handmaker, Sarah Mead and two anonymous reviewers for their helpful comments.

Disclosure statement

Both authors have served as advisers to the Government of Vanuatu at international climate change negotiations and consulted with members of the Government of Vanuatu about this article. However, the views expressed in this article are personal.

Funding

This work was supported by Nederlandse Organisatie voor Wetenschappelijk Onderzoek [grant number 016.Veni.195.342].

ORCID

Margaretha Wewerinke-Singh http://orcid.org/0000-0002-7782-1857
Diana Hinge Salili http://orcid.org/0000-0002-8906-0653

References

AOSIS. (1991). *Submission on Behalf of AOSIS: Draft Annex Relating to Article 23 (Insurance) for Inclusion in the Revised Single Text on Elements Relating to Mechanisms.* In Intergovernmental Negotiating Committee for a Framework Convention on Climate Change: Working Group II, Vanuatu, (A/AC.237/WG.II/Misc.13). Submitted by the Co-Chairmen of Working Group II, 4th session, Agenda Item 2(b), UN Doc A/AC.237/WG.II/CRP.8.

Ardern, J. (2018, April 18). The commonwealth can Kickstart a global offensive on climate change. *The Guardian.* Retrieved from https://www.theguardian.com/commentisfree/2018/apr/18/commonwealth-global-climate-change-new-zealanders

Australian Broadcasting Corporation. (2015). *Tropical cyclone Pam.* Retrieved from https://www.abc.net.au/news/2015-03-21/un-raises-vanuatu-cyclone-death-toll/6337816

Benjamin, L., Thomas, A., & Haynes, R. (2018). An 'Islands' COP'? Loss and damage at COP23. *Review of European, Comparative and International Environmental Law, 27,* 332–340.

Bodanksy, D. (2017). The role of the international court of justice in addressing climate change: Some preliminary reflections. *Arizona State Law Journal, 49,* 659–712.

Brown, R. (2013, March 4). The rising tide of climate change cases. *The Yale Globalist.* Retrieved from http://tyglobalist.org/in-the-magazine/theme/the-rising-tide-of-climate-change-cases/

Burger, M., & Wentz, J. (2018). Holding fossil fuel companies accountable for their contribution to climate change: Where does the law stand? *Bulletin of the Atomic Scientists, 74*(6), 397–403.

Burkett, M. (2013). A justice paradox: On climate change, Small Island developing states, and the quest for effective legal remedy. *University of Hawaii Law Review, 35*(2), 633–670.

Burkett, M. (2014). Loss and damage. *Climate Law, 4*(1-2), 119–130.

Calliari, E. (2018). Loss and damage: A critical discourse analysis of parties' positions in climate change negotiations. *Journal of Risk Research, 21*(6), 725–747.

Cho, R. (2016). *El Niño and global warming- what's the connection?* State of the Planet, Earth Institute, Columbia University. Retrieved from https://blogs.ei.columbia.edu/2016/02/02/el-nino-and-global-warming-whats-the-connection/

Darby, M. (2018, April 16). UK labour supports call for "climate damages tax" on companies. *Climate Home News*. Retrieved from http://www.climatechangenews.com/2018/04/16/uk-labour-supports-call-climate-damages-tax-oil-companies/

Duffy, H. (2018). *Strategic human rights litigation: Understanding and maximising impact.* Oxford: Hart Publishing.

Esler, S. (2015). *Vanuatu post-disaster needs assessment: Tropical cyclone Pam.* Assessment Report, Government of Vanuatu. Retrieved from http://dfat.gov.au/about-us/publications/Pages/vanuatu-post-disaster-needs-assessment-tropical-cyclone-pam-march-2015.aspx

Farber, D. A. (2008). The case for climate compensation: Justice for climate change victims in a complex world. *Utah Law Review, 1*, 377–413.

Frumhoff, P., Heede, R., & Oreskes, N. (2015). The climate responsibilities of industrial carbon producers. *Climatic Change, 132*(2), 157–171.

Gage, A., & Wewerinke, M. (2015). *Taking climate justice into our own hands: A model climate compensation act.* Vancouver: West Coast Environmental Law.

Ganguly, G., Setzer, J., & Heyvaert, V. (2018). If at first you don't succeed: suing corporations for climate change. *Oxford Journal of Legal Studies, 38*(1), 841–868.

Garschagen, M., et al. (2015). *World risk report 2015.* Report, United Nations University Institute for Environment and Human Security and Bündnis Entwicklung Hilft. Retrieved from https://collections.unu.edu/eserv/UNU:3303/WRR_2015_engl_online.pdf

Global Facility for Disaster Reduction and Recovery. (2009). *Reducing the risk of disasters and climate variability in the Pacific Islands.* Republic of Vanuatu Country Assessment, World Bank. Retrieved from https://www.sprep.org/att/irc/ecopies/countries/vanuatu/113.pdf

Government of Vanuatu. (2018a). *National policy on climate change and disaster-induced displacement.* Retrieved from https://www.iom.int/sites/default/files/press_release/file/iom-vanuatu-policy-climate-change-disaster-induced-displacement-2018.pdf

Government of Vanuatu. (2018b, December 3). *High-level statement of Hon. Minister Ralph Regenvanu to the 24th Conference of the parties to the UNFCCC.* On file with the authors.

Hafner-Burton, E. M., Victor, D. G., & Lupu, Y. (2012). Political science research on international law: The state of the field. *The American Journal of International Law, 106*(1), 47–97.

Heltberg, R., Siegel, P. B., & Jorgensen, S. L. (2009). Addressing human vulnerability to climate change: Toward a "no-regrets" approach. *Global Environmental Change, 19*(1), 89–99.

Hoffmaister, J. P., Talakai, M., Damptey, P., & Barbosa, A. S. (2014). *Warsaw international mechanism for loss and damage: Moving from polarizing discussions towards addressing the emerging challenges faced by developing countries.* Kuala Lumpur: Third World Network. Retrieved from http://www.twn.my/title2/climate/info.service/2014/cc140101/Opinion%20LD-060114.pdf

Hunter, D. B. (2009). The implications of climate change litigation: Litigation for international environmental law-making. In W. C. G. Burns & H. M. Osofsky (Eds.), *Adjudicating climate change: State, national and international approaches* (pp. 357–374). Cambridge: Cambridge University Press.

Huq, S., Roberts, E., & Fenton, A. (2013). Loss and damage. *Nature Climate Change, 3*, 947–949.

Hurley, L. (2011, September 2018). Island nation girds for legal battle against industrial emissions. *The New York Times*. Retrieved from https://archive.nytimes.com/www.nytimes.com/gwire/2011/09/28/28greenwire-island-nation-girds-for-legal-battle-against-i-60949.html?pagewanted=all

Ieyoub and Eisenberg. (2000). State attorney general actions, the tobacco litigation, and the doctrine of Parens Patriae. *Tulane Law Review, 74*, 1859–1884.

Intergovernmental Panel on Climate Change (2018). *Global warming of 1.5°C. An IPCC special report on the impacts of global warming of 1.5°C above pre-industrial levels and related global greenhouse gas emission pathways, in the context of strengthening the global response to the threat of climate change, sustainable development, and efforts to eradicate poverty.* Edited by V. Masson-Delmotte, et al. Geneva, Switzerland: World Meteorological Organization.

Isaac, J., Jumeau, R., Mahmud, A. I., & Regenvanu, R. (2018, May 2). When will the world's polluters start paying for the mess they made? *Climate Home News.* Retrieved from http://www.climatechangenews.com/2018/05/02/will-worlds-polluters-start-paying-mess-made

Jacobs, R. E. (2005). Treading deep waters: Substantive law issues in Tuvalu's threat to sue the United States in the international court of justice. *Pacific Rim Law & Policy Journal, 14*(1), 103–128.

Kyoto Protocol. (1997). 2303 UNTS 148.

Kysar, D. A. (2011). What climate change can do about tort law. *Environmental Law, 41*(1), 1–71.

Lees, E. (2017). Responsibility and liability for climate loss and damage after Paris. *Climate Policy, 17*(1), 59–70.

Linnerooth-Bayer, J., Mace, M. J., & Verheyen, R. (2003). *Insurance-related actions and risk assessment in the context of the UNFCCC.* Bonn UNFCCC Secretariat.

Mace, M. J., & Verheyen, R. (2016). Loss, damage and responsibility after COP21: All options open for the Paris agreement. *Review of European Community & International Environmental Law, 25*(2), 197–214.

Oxford Living Dictionary. (2018). *Compensation*. Retrieved from https://en.oxforddictionaries.com/definition/compensation

Pacific Catastrophe Risk Assessment and Financing Initiative. (2011). *Country risk profile: Vanuatu*. Retrieved from http://pcrafi.spc.int/documents/136

Paris Agreement. (2015).

Peel, J., & Osofsky, H. M. (2015). *Climate change litigation: Regulatory pathways to cleaner energy*. Cambridge: Cambridge University Press.

Pekkarinen, V., Toussaint, P., & Van Asselt, H. (2019). Loss and damage after Paris: Moving beyond rhetoric. *Carbon and Climate Law Review, 13*(1), 31–49.

Rajamani, L. (2015). Addressing loss and damage from climate change impacts. *Economics & Political Weekly, L, 30*, 17–21.

Regenvanu, R., & Persaud, A. (2018, December 10). It's time for those who caused climate change to pay for it. *Reuters*. Retrieved from http://news.trust.org/item/20181206165251-g7ixe/

Republic of Vanuatu. (2018a). *Submission to the executive committee of the Warsaw international mechanism for loss and damage of the UNFCCC*. Retrieved from http://unfccc.int/files/adaptation/workstreams/loss_and_damage/application/pdf/vanuatu_submission.pdf

Republic of Vanuatu. (2018b). *Declaration by the Republic of Vanuatu*. United Nations Treaty Collection. Retrieved from https://treaties.un.org/Pages/ViewDetails.aspx?src=TREATY&mtdsg_no=XXVII-7-d&chapter=27&lang=_en&clang=_en

Republic of Vanuatu. (2018c). *Statement by the Honorable Ralph Regenvanu, Minister of Foreign Affairs, International Cooperation and external trade of the Republic of Vanuatu at the 2018 Climate Vulnerable Forum virtual summit on 22 November 2018*. Retrieved from https://www.youtube.com/watch?v=kst10ZfSKPc

Reuters. (2002, August). Tuvalu seeks help in US global warming Lawsuit. Retrieved from https://www.open.edu/openlearn/ocw/pluginfile.php/619493/mod_resource/content/1/reading1d.pdf

Richards, J.-A., Hillman, D., & Boughey, L. (2018, December). *The climate damages tax: A guide to what it is and how it works*. London: Stamp Out Poverty. Retrieved from https://www.stampoutpoverty.org/cdt/further-information/

Roberts, E., & Huq, S. (2015). Coming full circle: The history of loss and damage under the UNFCCC. *International Journal of Global Warming, 8*(2), 141–157.

Sands, P. (2016). Climate change and the rule of law: Adjudicating the future in international law. *Journal of Environmental Law, 28*, 19–35.

Savaresi, A., Hartmann, J., & Cismas, I. (2019). *The impacts of climate change and human rights: Some early reflections on the carbon majors inquiry*. Retrieved from https://papers.ssrn.com/sol3/papers.cfm?abstract_id=3277568

Schwarte, C., & Byrne, R. (2011). *International climate change litigation and the negotiation process*. London: FIELD Working Paper. Retrieved from http://www.indiaenvironmentportal.org.in/files/FIELD_cclit_long_Oct.pdf

Siegele, L. (2018). Loss and damage (Article 8). In D. Klein, M. P. Carazo, M. Doelle, J. Bulmer, & A. Higham (Eds.), *The Paris agreement on climate change: Analysis and commentary* (pp. 224–238). Oxford: Oxford University Press.

Singh, H. (2018, May 14). *Expert dialogue underscores finance gap to address climate induced "loss and damage"*. Kuala Lumpur: Third World Network. Retrieved from https://twnetwork.org/climate-change/expert-dialogue-underscores-finance-gap-address-climate-induced-loss-and-damage

Sopko, S. P., & Falvey, R. J. (2015). *Annual tropical cyclone report 2015*. Joint Typhoon Warning Center.

SPREP. (2015). *We the Pacific: Amplifying the Pacific voice at COP21*. Press Release of 2 December 2015. Retrieved from http://www.scoop.co.nz/stories/WO1512/S00009/we-the-pacific-amplifying-the-pacific-voice-at-cop21.htm

Tschakert, P., Ellis, N. R., Anderson, C., Kelly, A., & Obeng, J. (2019). One thousand ways to experience loss: A systematic analysis of climate-related intangible harm from around the world. *Global Environmental Change, 55*, 58–72.

UN Department of Economic and Social Affairs. (2010). *Vanuatu national assessment report: 5 year review of the Mauritius strategy for further implementation of the Barbados programme of action for sustainable development*. New York: United Nations.

UNFCCC. (2007). *Bali action plan*. Decision 1/CP.13, FCCC/CP/2007/6/Add.1.

UNFCCC. (2013). *Warsaw international mechanism for loss and damage associated with climate change impacts*. Decision 2/CP.19, FCCC/CP/2013/10/Add.1.

UNFCCC. (2014). *Report of the executive committee of the Warsaw international mechanism for loss and damage associated with climate change impacts*. Subsidiary Body for Scientific and Technological Advice, FCCC/SB/2014/4, Annex II.

UNFCCC. (2015a). *Adoption of the Paris agreement*. Decision 1/CP.21, FCCC/CP/2015/10/Add.1.

UNFCCC. (2015b). *Draft Paris outcome: Proposal by the president*. Draft Text on COP21 agenda item 4(b), Durban Platform for Enhanced Action (decision 1/CP.17) Adoption of a Protocol, Another Legal Instrument, or an Agreed Outcome with Legal Force under the Convention Applicable to All Parties, Version 2 of 10 December 2015 at 21:00 (on file with the authors).

UNFCCC. (2017a). *Report of the executive committee of the Warsaw international mechanism for loss and damage associated with climate change Impacts*. Subsidiary Body for Scientific and Technological Advice, Doc FCCC/SB/2017/1/Add1, Annex.

UNFCCC. (2017b). *Warsaw international mechanism for loss and damage associated with climate change impacts*. Decision 5/CP.23, FCCC/CP/2017/11/Add.1.

UNFCCC. (2018a). *Report of the Suva expert dialogue on loss and damage associated with climate change impacts*. Bonn: UNFCCC Secretariat. Retrieved from https://unfccc.int/sites/default/files/resource/SUVA%20Report_ver_20180914.pdf

UNFCCC. (2018b). *Report of the executive committee of the Warsaw international mechanism for loss and damage associated with climate change impacts.* Advance unedited version. Retrieved from https://unfccc.int/sites/default/files/resource/cp24_auv_ec%20wim.pdf

UN Framework Convention on Climate Change (UNFCCC). (1992). 1771 UNTS 107.

United Kingdom Sustainable Investment and Finance Association and Climate Change Collaboration. (2018). *Not long now: survey of fund managers' responses to climate-related risks facing fossil fuel companies.* Retrieved from http://uksif.org/wp-content/uploads/2018/04/UPDATED-UKSIF-Not-Long-Now-Survey-report-2018-ilovepdf-compressed.pdf

UN Office of the High Representative for the Least Developed Countries, Landlocked Developing Countries and Small Island Developing States. (2018). *Graduation from LDC category: Historical background.* New York: UN DESA. Retrieved from http://unohrlls.org/custom-content/uploads/2018/12/historical-background-on-graduation-updated-Dec2018.pdf

Vanhala, L., & Hestbaek, C. (2016). Framing climate change loss and damage in UNFCCC negotiations. *Global Environmental Politics, 16*(4), 111–129.

Vanuatu Metrology and Geo-Hazards Department. (2017). *Increasing resilience to climate change and natural hazards: environment and social management framework.* World Bank Project Doc No SFG2255 REV, Government of Vanuatu. Retrieved from http://documents.worldbank.org/curated/en/349571468132625631/pdf/SFG2255-REVISED-EA-P112611-Box402910B-PUBLIC-Disclosed-5-23-2017.pdf

Vanuatu National Statistics Office. (2017). *2016 post-TC Pam mini-census report.* Port Vila: Government of Vanuatu. Retrieved from https://vnso.gov.vu/index.php/component/advlisting/?view=download&fileId=4542

Verheyen, R. (2005). *Climate change damage and international law: Prevention duties and state responsibility.* Leiden: Brill.

Wallimann-Helmer, I. (2015). Justice for climate loss and damage. *Climatic Change, 133*(3), 469–480.

Warner, K., & Zakieldeen, S. A. (2011). *Loss and damage due to climate change: An overview of the UNFCCC negotiations.* European Capacity Building Initiative. Retrieved from https://oxfordclimatepolicy.org/publications/documents/LossandDamage.pdf

Wewerinke-Singh, M. (2019). *State responsibility, climate change and human rights under international law.* Oxford: Hart Publishing.

Wewerinke-Singh, M., & Doebbler, C. (2016). The Paris agreement: Some critical reflections on process and substance. *University of New South Wales Law Journal, 39*(4), 1486–1517.

Wewerinke-Singh, M., & Van Geelen, T. (2019). Protection of climate displaced persons under international law: A case study from Mataso Island, Vanuatu. *Melbourne Journal of International Law, 19*(2), 1–37.

OPEN ACCESS

Interpreting the UNFCCC's provisions on 'mitigation' and 'adaptation' in light of the Paris Agreement's provision on 'loss and damage'

Morten Broberg

ABSTRACT

This article examines how the introduction of a specific provision on loss and damage (L&D) in the Paris Agreement affects the construction of provisions on 'mitigation' and 'adaptation' as established within the treaty framework of the United Nations Framework Convention on Climate Change (UNFCCC). It shows that the establishment of L&D at treaty level has created a legal basis for finding 'responsibility' for adverse consequences that can be attributed to the failure to fulfil UNFCCC obligations as laid down in the provisions on mitigation and adaptation. This, it argues, strengthens the legal basis for pursuing remedies aimed at reparation for these consequences, such as the establishment of climate change funds and of insurance solutions. Moreover, it demonstrates that prior to establishing L&D at treaty level, L&D issues and measures (such as the Warsaw International Mechanism) were treated in legal terms within the framework of adaptation. However, with the adoption of the Paris Agreement, L&D has been given its own legal basis and therefore L&D issues and measures must henceforth be treated within this new framework.

Key policy insights

- The provisions on mitigation and adaptation in the UNFCCC may be re-interpreted in light of Article 8 of the Paris Agreement so as to instil a legal responsibility in the UNFCCC provisions on mitigation and adaptation.
- The limits of 'adaptation' must be re-defined in light of the adoption of the Paris Agreement and the introduction of L&D as a third pillar of international climate change law.

Introduction

Amongst the most contentious issues during the negotiations that led to the adoption of the Paris Agreement was the introduction of a dedicated provision on loss and damage (L&D) in Article 8. Thus, several developing countries were fervent supporters of introducing L&D in the Paris Agreement whilst many developed countries were strongly opposed to this. Whereas Article 8 has been the subject of much analysis, it appears that no-one has examined how it may affect the construction of provisions on 'mitigation' and 'adaptation' as established within the treaty framework of the United Nations Framework Convention on Climate Change (UNFCCC). This article provides such examination. It argues that the establishment of L&D at treaty level has created a legal basis for finding 'responsibility' for adverse consequences that can be attributed to the failure to fulfil UNFCCC obligations as laid down in the provisions on mitigation and adaptation. This, the article argues, strengthens the legal basis for pursuing remedies aimed at reparation for those consequences, such as the establishment of climate change funds and of insurance solutions. Moreover, the article demonstrates that

This is an Open Access article distributed under the terms of the Creative Commons Attribution-NonCommercial-NoDerivatives License (http://creativecommons.org/licenses/by-nc-nd/4.0/), which permits non-commercial re-use, distribution, and reproduction in any medium, provided the original work is properly cited, and is not altered, transformed, or built upon in any way.

prior to establishing L&D at treaty level, L&D issues and measures (such as the Warsaw International Mechanism; Decision 2/CP.19, 2014) were treated in legal terms within the framework of adaptation. However, with the adoption of the Paris Agreement, L&D has been given its own legal basis, meaning that L&D issues must now be treated within this new framework. Finally, the introduction of L&D was based upon a compromise deal; on the one hand, L&D was established in the Paris Agreement itself. On the other hand, important limitations on the contents of L&D were laid down in paragraph 51 of decision 1/CP.21 whereby the Paris Agreement was adopted. Against this background the article observes that it is much easier to amend decision 1/CP.21 than it is to amend the Paris Agreement as such. Thus, in principle it is possible to abandon the present limitations on the contents of L&D.

The three pillars of international climate change law

With the 1992 adoption of the United Nations Framework Convention on Climate Change (UNFCCC), 'mitigation' was established as the first pillar of international climate change law, with 'adaptation' as its second. The subsequent 2015 Paris Agreement added a third pillar; namely 'loss and damage' or 'L&D'. Normally, mitigation, adaptation, and L&D are presented as being closely inter-related and mutually complementary, as shown in Figure 1. Metaphorically, climate change may be viewed as an enemy against which mitigation, adaptation, and L&D form an integrated fortress. Mitigation constitutes the outermost rampart by curbing greenhouse gas emissions.[1] However, the enemy has broken through this outermost layer of protection, and so adaptation to the climate changes that are happening is required – adaptation, therefore, makes up the intermediate rampart.[2] Finally, in those situations where adaptation is insufficient, we will have to rely upon L&D, the inner rampart.[3]

When viewing L&D as a new strategy to complement the two original 'ramparts' which stave off the threats posed by climate change, it seems only natural to focus upon those challenges that mitigation and adaptation have been unable to meet, but which L&D is supposedly able to address, as well as those which are left unmet by either mitigation, adaptation, or L&D.

Obviously, the description of mitigation, adaptation, and L&D as an integrated fortress is nothing but a metaphor. In reality, the three 'ramparts' originate from provisions in legally binding international agreements. L&D is therefore an addition to an existing legal web and, rather than pursuing a narrative where L&D merely complements mitigation and adaptation, we must consequently take a legal approach to the three concepts viewed together. The establishment of L&D at treaty level as a third pillar of international climate change law means that we must henceforth interpret the two original pillars – mitigation and adaptation – in the light of the new L&D pillar.

Put differently, *the inclusion of L&D in the Paris Agreement has the potential to amend the legal contents of mitigation and adaptation*. This amendment may happen, firstly, through a substantive change to the fundamental interpretation of the provisions on mitigation and adaptation and, secondly, by L&D taking over some of the substantive coverage of one or both of the two old pillars. Below we shall examine these two possible impacts of L&D upon mitigation and adaptation.

Changing the interpretation of the provisions on 'mitigation' and 'adaptation'

According to the principles on legal interpretation of international agreements, the advent of a new treaty provision, such as the one on L&D in the Paris Agreement, has the potential to change our understanding of the legal contents of existing concepts, such as those on mitigation and adaptation. When interpreting an

Figure 1. The three pillars of international climate change law.

international agreement such as the 1992 UNFCCC or the 2015 Paris Agreement, lawyers will turn to the 1969 Vienna Convention on the Law of Treaties which provides a codification of customary international law and state practice concerning treaties. According to the Convention's Article 31(1), '[a] treaty shall be interpreted in good faith in accordance with the ordinary meaning to be given to the terms of the treaty ... '. Thus, the ordinary meaning of a given treaty provision constitutes the interpretative starting point. However, Article 31(1) goes on to state that this meaning of the treaty terms must be derived 'in their context and in the light of [the treaty's] object and purpose'. In Article 31(2)-(4), the Vienna Convention elaborates upon Article 31(1). In this respect it is important to observe that Article 31(3) further states that *subsequent* agreements and practices may also have to be taken into account for the purposes of interpreting a treaty provision.

The above principles mean that, when interpreting the provisions on 'mitigation' and 'adaptation' after the entry into force of the Paris Agreement, it is necessary to take into due account the fact that Article 8 of the Paris Agreement now establishes L&D as a (new) third pillar of international climate change law. And this applies both to those parts of Article 8 which are legally binding (applying the term 'shall') and those that rather denote a recommendation (applying the term 'should').

In order to determine what this re-interpretation entails, we must first clarify what the advent of Article 8 on L&D adds to the international climate change regulatory scheme more generally. To do so, our starting point must necessarily be the meaning of the terms 'loss' and 'damage'. Under international law these terms denote responsibility for harm – as, for example, reflected in Article 36 of the ILC Articles of 2001 on Responsibility of States for Internationally Wrongful Acts (ARSIWA). Or, in the words of Crawford (2019), '"[d]amage" denotes loss, *damnum,* usually a financial quantification of physical or economic injury or damage or of other consequences of [a breach of an international obligation]'. Thus, in an international law context, the terms 'loss and damage' inevitably evoke the notions of 'responsibility' and 'reparation'.

For the purposes of our interpretation of Article 8 of the Paris Agreement, three terms play particularly important roles: 'responsibility', 'liability' and 'compensation': Whereas the notion of (state) *responsibility* is well established as referring to acts which are unlawful under international law, the term *liability* is much less clearly defined. As pointed out by Crawford (2013) as well as by Brown and Seck (2013), the term has been used to distinguish it from 'responsibility' – for example to cover situations of transboundary harm caused by activities that prima facie are not unlawful under international law. If a state commits an internationally wrongful act, this may entitle others to respond; by seeking cessation and/or reparation. Reparation may take the form of restitution in kind, *compensation* through the payment of money and satisfaction which covers the residual ways of redressing the wrong (Crawford, 2019).

Thus, whereas 'responsibility' is well established under international law and comes with a distinct set of consequences, 'liability' is not clearly defined. However, where drafters of a legal text use both of these terms without indicating that the two terms are meant to be synonymous, we may reasonably assume that 'liability' covers something else. Finally, 'compensation' is a subset to reparation.

Within the climate change debate, the question of south–north relations and climate justice has played – and continues to play – a key role (Mickelson, 2009). This is also true with regard to the introduction of L&D which constitutes a compromise between south and north (Calliari et al., 2019). Thus, the first discussions on the introduction of a scheme on L&D in the field of climate change law were initiated by the Alliance of Small Island States (AOSIS) during the negotiations that led to the adoption of the UNFCCC in 1992. AOSIS argued that it was necessary to link climate change mitigation to a compensatory scheme and therefore proposed the establishment of an 'Insurance Pool', whose resources 'should be used to compensate' the most vulnerable developing countries for 'loss and damage' resulting from sea level rise. The financial burden of doing so would be 'distributed in an equitable manner' amongst the developed countries (Intergovernmental Negotiating Committee for a Framework Convention on Climate Change (INC), 1991). The fact that L&D was only established at treaty level with the Paris Agreement in 2015 may primarily be attributed to the reluctance of developed countries to accept responsibility that could ultimately result in liability to pay damages. This is evidenced in that, prior to the Paris Agreement negotiations, certain developed countries expressed concerns that establishing L&D at treaty level could provide a basis for holding them financially accountable. For example, in an interview just before the Paris Climate Conference, the serving US Secretary of State, John Kerry, explained that the negotiations on L&D would be particularly complicated. Whilst not denying that there was a degree of responsibility, Kerry made it very clear that the US would not support the introduction of L&D at treaty level, if it could be framed in such a way as to create a legal remedy (Goodell, 2015). Indeed, the Paris Agreement's end result was to introduce L&D in Article 8 of the Agreement, while in connection with the Agreement's formal adoption in decision 1/CP.21 the parties explicitly stated 'that Article 8 of the Agreement does not involve or provide a basis for any liability or compensation'.

From the above it follows that Article 8 of the Paris Agreement introduces obligations upon the Parties to the Agreement 'with respect to loss and damage associated with the adverse effects of climate change'. However, whereas Parties failing to comply with these obligations may incur *responsibility* under the Agreement, the provision as such does not provide a basis for *liability* or *compensation*.

Similarly, Lees (2017) has pointed out that the Paris Agreement distinguishes between 'responsibility' and 'liability', and that paragraph 51 of decision 1/CP.21 excludes direct claims for compensation, or other kinds of liability, arising from loss or damage responsibility. In other words, like the present author, Lees takes the view that the Paris Agreement recognizes 'responsibility' under international law for climate change induced L&D. Along the same lines, Voigt (in press), has observed that paragraph 51 of decision 1/CP.21 does 'not prevent the application of general rules of international law on state responsibility and liability for damage resulting from the breach of other treaties and/or customary norms'.

The decisive question is, therefore, what the consequences of this finding are, since the Paris Agreement itself does not provide a suitable legal remedy for pursuing compensation from those responsible for causing L&D. In this respect, Doelle and Seck (2019) have suggested that it may be possible to find a legal basis for pursuing L&D claims outside the Paris Agreement's legal framework, including on the basis of both domestic laws and international legal systems. Building on Doelle and Seck's observations, finding that Article 8 establishes a 'general principle of responsibility for L&D' strengthens the case to claim reparation by referring to these other legal bases. Consequently, even if Article 8 of the Paris Agreement merely establishes a principle of responsibility (but precludes liability and compensation) this may have an impact upon whether claimants can potentially obtain some kind of reparation on the basis of remedies outside the Paris Agreement.

From the above it follows that the introduction of L&D in Article 8 of the Paris Agreement entails a legal strengthening of the duties which the parties to the Agreement assume through the provisions on 'mitigation' and 'adaptation'. Since L&D concerns those climate change impacts that are attributable to the 'residual' of mitigation and adaptation, then there is an undeniable logical connection between L&D, on the one hand, and mitigation and adaptation on the other. Even though the Paris Agreement does not provide a basis for finding 'liability' and 'compensation', the fact that it provides a legal basis for finding 'responsibility' allows us to consider other remedies including those that envisage reparation, such as the establishment of climate change funds (as proposed by Brown & Seck, 2013) and insurance solutions. Since L&D is the residual of mitigation and adaptation, when deciding who is required to cover the costs of the L&D measures, it would seem natural to turn first to those parties who have assumed commitments in the field of mitigation and adaptation which have not been fully met. *In other words, based on the above, I suggest that it is possible to interpret the provisions on mitigation and adaptation in light of Article 8 of the Paris Agreement so as to instil a legal responsibility in the aforementioned provisions.*

New delimitation of 'adaptation'

'Mitigation', 'adaptation', and 'L&D' perform different tasks within international climate change law. Not only do they pursue different objectives, they also apply different instruments, have separate bodies to perform different tasks within the remit of each of the three pillars, apply different procedures, and differ significantly when it comes to funding mechanisms and funding streams (UNFCCC, 2020). Consequently, whether an issue or measure must be categorized as mitigation, adaptation, or L&D may be of material importance.

When the Paris Agreement employs 'mitigation', 'adaptation', and 'L&D' as the three central pillars to pursue diverging objectives based on different mechanisms, it is necessary to distinguish clearly between them. To this end, we must first refer back to the terms and objectives of the relevant provisions. In this respect, the distinction between 'mitigation' and 'adaptation' has been fairly clear since the adoption of the UNFCCC in 1992. In contrast, there is considerable overlap between the concepts of 'adaptation' and 'L&D'. This is clearly reflected in the

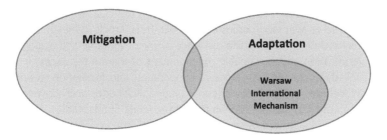

Figure 2. Mitigation, adaptation and the WIM before the adoption of the Paris Agreement.

establishment of the so-called 'Warsaw International Mechanism for Loss and Damage associated with Climate Change Impacts (Loss and Damage Mechanism)' (WIM) at the Warsaw COP19 in 2013. According to paragraph 1 of the WIM, the Mechanism was established by the UNFCCC 'under the Cancun Adaptation Framework', and intended as a workstream in the strategy of 'Adaptation and Resilience'.[4] Thus, before the adoption of the Paris Agreement, L&D was to be viewed as part of 'adaptation' (Figure 2). After the adoption of the Paris Agreement, however, it has become apparent that, from a strictly legal perspective, we now must distinguish the two. Therefore, those issues that fall within the scope of the WIM, in principle, no longer fit into the adaptation pillar, but instead belong under the pillar of L&D (Figure 3).

While it is theoretically quite straightforward to claim that, following the adoption of the Paris Agreement and the consequent establishment of L&D as a third pillar of international climate change law, issues that may be identified as L&D should no longer be categorized as 'adaptation', it is much more difficult to determine the distinction between adaptation and L&D in practice. According to Mayer (2018) this is due, in large part, to the fact that '[d]eveloped States remain generally reluctant to recognize loss and damage as a distinct type of response to climate change, seemingly due to fears of the claims for compensation or, otherwise, for the support that this would fuel'. Another arguable reason is that a number of measures which are explicitly listed in Article 8 on L&D of the Paris Agreement would have indisputably been considered to be part of adaptation prior to 2015 (Bodansky et al., 2017). By way of illustration, when a number of regional multi-country insurance risk pools were established in the Global South to address certain weather-related hazards, such as hurricanes, before the introduction of the Paris Agreement, they were seen to be measures of climate change adaptation (Broberg & Hovani-Bue, 2019; McAneney et al., 2013). Today, however, it would effectively be more correct to categorize these insurance schemes as L&D (Broberg, 2019).

As explained above, there is an apparent reluctance to delimit the issues and measures that must be categorized as L&D rather than adaptation, meaning that the boundaries between the two pillars are still rather blurry. Thus, whereas it would seem that the Paris Agreement allows us to categorize physical measures (e.g. building dykes against rising sea level) as adaptation, but identifies insurance solutions as L&D, the Agreement is much less

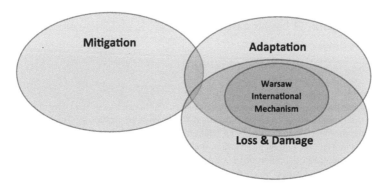

Figure 3. Mitigation, adaptation, loss & damage and the WIM after the adoption of the Paris Agreement.

clear when it comes to other types of measures. For example, it is not clear when efforts to make communities *resilient* should be considered as adaptation or instead as L&D. Thus, Article 7(9)(e) of the Paris Agreement stipulates that adaptation may include '[b]uilding the resilience of socioeconomic and ecological systems, including through economic diversification and sustainable management of natural resources', whereas paragraph 4(h) of Article 8 on L&D similarly concedes that 'areas of cooperation and facilitation to enhance understanding, action and support may include … [r]esilience of communities, livelihoods and ecosystems'.

Hence, the introduction of L&D means a redefinition of the limits of adaptation. Whereas this will affect the practical framework for the enforcement of some of those rights and obligations that have been established within the field of adaptation, it does not entail that these rights/obligations cease to apply.

Although it may be challenging to draw a clear distinction between adaptation and L&D, it is nevertheless clear that, *with the adoption of the Paris Agreement, the limits of 'adaptation' must be defined in light of the introduction of L&D as a third pillar of international climate change law.*

Conclusions

The introduction of L&D as a third pillar of international climate change law not only affects the notion of L&D itself, but also has consequences for the two pre-existing pillars, namely, mitigation and adaptation.

Firstly, L&D provides a legal basis for finding 'responsibility' and thereby engenders a legal strengthening of the duties which the parties to the Paris Agreement assume within the fields of 'mitigation' and 'adaptation'. This affects the very content and interpretation of the provisions on mitigation and adaptation, and, in particular, is likely to strengthen the legal basis for pursuing remedies aimed at reparation, such as supporting climate change funds and insurance solutions.

Secondly, prior to the adoption of the Paris Agreement, L&D was essentially viewed as part of adaptation. However, with the establishment of L&D as a third pillar of international climate change law, L&D issues shall no longer be treated as matters of adaptation, essentially meaning a material reduction of the scope of adaptation.

Taking a wider perspective, the above findings give rise to a number of questions that merit further research. First of all, what are the practical implications of the fact that the L&D provision in the Paris Agreement strengthens the legal basis for pursuing remedies aimed at reparation; including, what are the exact obligations that would arguably be breached, and when would the breach occur?

Also, taking a less legal approach, it may be useful to recall that the USA worked hard in the negotiation phase to ensure that the Paris Agreement's provision on L&D would not establish a legal basis for liability and compensation. This 'limitation' on L&D was introduced in paragraph 51 of the decision adopting the Agreement; not in the Agreement itself. Therefore, the Parties may revoke the limitation by having the COP adopt a new decision. This does not mean that we may expect the adoption of any new decision revoking or otherwise amending paragraph 51 to be straightforward; it is merely to say that in practice, adopting such a new decision would be appreciably easier than making a change to the Paris Agreement as such (Mace & Verheyen, 2016). With President Trump having decided that the USA will withdraw from the Paris Agreement, the possibility of such revocation of paragraph 51 has increased.

Notes

1. Mitigation refers to human intervention to reduce the sources or enhance the sinks of greenhouse gases.
2. In 2014, the Intergovernmental Panel on Climate Change (IPCC) defined adaptation as: 'The process of adjustment to actual or expected climate and its effects. In human systems, adaptation seeks to moderate or avoid harm or exploit beneficial opportunities. In some natural systems, human intervention may facilitate adjustment to expected climate and its effects.'
3. There is no universally accepted definition of L&D. For our purposes, L&D will be defined as covering those measures that address the impacts of climate change which are 'residual' to mitigation and adaptation.
4. Even in early 2020, the UNFCCC's website places the WIM as a workstream under 'Adaptation and resilience', cf. https://unfccc int/topics#:28deb0e5-7301-4c3f-a21f-8f3df254f2a4:33c8e50f-9dde-4fe5-9524-1949dd3f5bdb (last visited 13 January 2020).

Acknowledgements

The author would like to express his gratitude for excellent comments and suggestions provided by three anonymous reviewers. The usual waiver applies.

Disclosure statement

No potential conflict of interest was reported by the author(s).

References

Bodansky, D., Brunnée, J., & Rajamani, L. (2017). *International climate change Law*. Oxford University Press.

Broberg, M. (2019). Parametric loss and damage insurance schemes as a means to enhance climate change resilience in developing countries. *Climate Policy*. https://doi.org/10.1080/14693062.2019.1641461

Broberg, M., & Hovani-Bue, E. (2019). Disaster risk reduction through risk pooling: The case of hazard risk pooling schemes. In K. L. H. Samuel, M. Aronsson-Storrier, & K. Nakjavani Bookmiller (Eds.), *The Cambridge handbook of disaster risk reduction and international law* (pp. 257–274). Cambridge University Press.

Brown, C., & Seck, S. (2013). Insurance law principles in an international context: Compensating losses caused by climate change. *Alberta Law Review*, *50*(3), 541–576. https://doi.org/10.29173/alr96

Calliari, E., Surminski, S., & Mysiak, J. (2019). The politics of (and behind) the UNFCCC's loss and damage mechanism. In R. Mechler, L. M. Bouwer, T. Schinko, S. Surminski, & J. Linneroth-Bayer (Eds.), *Loss and damage from climate change – concepts, methods and policy options* (pp. 155–178). Springer Open.

Crawford, J. (2013). *State responsibility: The general part*. Cambridge University Press.

Crawford, J. (2019). *Brownlie's principles of public international Law*. Oxford University Press.

Doelle, M., & Seck, S. (2019). Loss & damage from climate change: From concept to remedy? *Climate Policy*, https://doi.org/10.1080/14693062.2019.1630353

Goodell, J. (2015, December 1). John Kerry on climate change: The fight of our time. *Rolling Stone*. https://www.rollingstone.com/culture/culture-news/john-kerry-on-climate-change-the-fight-of-our-time-50220/#ixzz3tAfuZPV5

ILC Articles. (2001). Responsibility of states for internationally wrongful acts 2001. Text adopted by the Commission at its fifty-third session, in 2001, and submitted to the General Assembly as a part of the Commission's report covering the work of that session. *Official Records of the General Assembly, Fifty-sixth Session, Supplement No. 10 (A/56/10)*.

Intergovernmental Negotiating Committee for a Framework Convention on Climate Change (INC). (1991). Vanuatu: Draft annex relating to Article 23 (Insurance) for inclusion in the revised single text on elements relating to mechanisms (A/AC.237/WG.II/Misc.13) submitted by the Co-Chairmen of Working Group II.

Intergovernmental Panel on Climate Change. (2014). Summary for policymakers. In C. B. Field, V. R. Barros, D. J. Dokken, K. J. Mach, M. D. Mastrandrea, T. E. Bilir, M. Chatterjee, K. L. Ebi, Y. O. Estrada, R. C. Genova, B. Girma, E. S. Kissel, A. N. Levy, S. MacCracken, P. R. Mastrandrea, & L. L. White (Eds.), *Climate change 2014: Impacts, adaptation, and vulnerability. Part A: Global and sectoral aspects. Contribution of Working Group II to the Fifth Assessment Report of the Intergovernmental Panel on climate change*. Cambridge University Press. https://www.ipcc.ch/site/assets/uploads/2018/02/ar5_wgII_spm_en.pdf

Lees, E. (2017). Responsibility and liability for climate loss and damage after Paris. *Climate Policy*, *17*(1), 59–70. https://doi.org/10.1080/14693062.2016.1197095

Mace, M. J., & Verheyen, R. (2016). Loss, damage and responsibility after COP21: All options open for the Paris Agreement. *Review of European Community & International Environmental Law*, *25*(2), 197–214. https://doi.org/10.1111/reel.12172

Mayer, B. (2018). *The international law on climate change*. Cambridge University Press.

McAneney, J., Crompton, R., McAneney, D., Musulin, R., Walker, G., & Pielke Jr, R. (2013). *Market-based mechanisms for climate change adaptation: Assessing the potential for and limits to insurance and market based mechanisms for encouraging climate change adaptation*. National Climate Change Adaptation Research Facility, Gold Coast – Australia.

Mickelson, K. (2009). Beyond a politics of the possible? South-north relations and climate justice. *Melbourne Journal of International Law*, *10*, 411–423. https://law.unimelb.edu.au/mjil/issues/issue-archive/102 and https://scholar.google.com/scholar_lookup?hl=en&publication_year=2009&pages=411-423&issue=2&author=Karin+Mickelson&title=Beyond+a+politics+of+the+possible%3F+South%E2%80%93North+relations+and+climate+justice

UNFCCC. (2020). *Introduction to climate finance – website*. Retrieved January 27, 2020, from https://unfccc.int/topics/climate-finance/the-big-picture/introduction-to-climate-finance

United Nations. (2014). *Decision 2/ CP.19, Warsaw international mechanism for loss and damage associated with climate change impacts*.

Vienna Convention on the Law of Treaties 1969, done at Vienna on 23 May 1969. Entered into force on 27 January 1980. United Nations, Treaty Series, vol. 1155, p. 331.

Voigt, C. (in press). International responsibility and liability. In L. Rajamani & J. Peel (Eds.), *Oxford handbook of international environmental law*. Oxford University Press.

A human rights-based approach to loss and damage under the climate change regime

Patrick Toussaint ⓘ and Adrian Martínez Blanco ⓘ

ABSTRACT

Climate change has been labelled the human rights challenge of the twenty-first century. Loss and damage resulting from climate change, in particular, poses a severe threat to the human rights of affected communities. However, the international response to climate change under the United Nations Framework Convention on Climate Change (UNFCCC) has thus far insufficiently taken human rights into account, contributing to policy outcomes inadequate to protecting communities affected by loss and damage. This article proposes the adoption of a human rights-based approach as a strategic tool for policymakers to strengthen the international response to loss and damage. The approach builds on the existing obligations of Parties under international and regional human rights treaties and provides a method for systematically integrating human rights that goes beyond mere mainstreaming of human rights. Specifically, the article identifies opportunities for anchoring such an approach under the Warsaw International Mechanism and key mechanisms for the implementation of the Paris Agreement. Conversely, it considers the integration of loss and damage in the work of relevant human rights bodies, specifically the United Nations Human Rights Council and the Office of the High Commissioner for Human Rights.

Key policy insights
- Adopting a human rights-based approach can be an important strategic tool for policymakers to strengthen the international response on loss and damage.
- Although the Paris Rulebook is weak on human rights, Parties are bound by their existing obligations under international and regional human rights treaties they have ratified and should be guided by the Paris Agreement's preamble.
- The Paris Rulebook sidelines Article 8 of the Paris Agreement, but loss and damage still plays an important role in the Transparency Framework and Global Stocktake.
- There is a significant opportunity for the Warsaw International Mechanism's Executive Committee to develop human rights guidelines for loss and damage policies and actions, as well as guidelines for conducting human rights impact assessments, and to set up a specialized body to monitor compliance.

1. Introduction

Current collective climate mitigation efforts are likely to result in temperatures exceeding the well below 2°C goal set out by the Paris Agreement (Rogelj et al., 2016). Despite adaptation efforts, some adverse impacts of climate change either are not or cannot be avoided (Verheyen & Roderick, 2008). Already today, loss and damage due to climate change impacts is leading to the impairment of a wide range of internationally

This article has been republished with minor changes. These changes do not impact the academic content of the article.

recognized human rights. Nonetheless, the current international response to loss and damage insufficiently takes human rights into account, resulting in policy outcomes that are inadequate to protect affected communities. There is thus a clear need for stronger integration of human rights and climate governance frameworks to address the human rights implications of loss and damage.

This article seeks to make a contribution by proposing to extend the application of human rights-based approaches (HRBAs) to loss and damage due to climate change. It argues that the adoption of an HRBA would present an opportunity to make headway in the policy debate on loss and damage under the climate change regime, in the face of political disagreements over liability and compensation (Lees, 2017). An HRBA can be understood as a method to integrate the normative dimension of human rights into the design, implementation, and evaluation of policies and actions. Knur (2014, p. 55) relevantly notes that 'the United Nations, as well as the human rights institutions, have not yet fully explored the advantages of a human rights-based approach to climate change'. Considering the variety of HRBAs advocated in different international policy fields, a tailored and strategic approach addressing loss and damage under the climate change regime could help integrate human rights considerations in a systematic way that goes beyond a mere rhetorical mainstreaming of human rights language into treaty texts. Due to its emphasis on the human rights of affected individuals and its special consideration of marginalized groups, an HRBA presents a unique opportunity to bring into the climate change process the voices of those most affected by loss and damage that have gone hitherto largely unheard (Toussaint, 2019). Thus, an HRBA has the potential to significantly strengthen the international response to climate change loss and damage while being anchored in the normative framework of human rights law.

This article begins by exploring the human rights implications of loss and damage. It then examines the extent to which human rights have entered the international policy discourse on loss and damage. Building on this analysis, it lays out the rationale for adopting an HRBA to loss and damage from climate change. Finally, the article identifies opportunities for the adoption of an HRBA in the international policy response to loss and damage specifically under the climate change regime, including under the Warsaw International Mechanism (WIM) and key mechanisms of the Paris Agreement.

2. The human rights implications of loss and damage

Before moving on to a discussion of the relevance of an HRBA to loss and damage, it is essential to examine the implications of loss and damage for human rights. As climate change intensifies, there is little hope that current mitigation and adaptation efforts will suffice to prevent both short- and long-term impacts. Although there is no universally accepted definition (IPCC, 2018, p. 454, Cross-Chapter Box 12; Mechler et al., 2019, p. 11), loss and damage is commonly referred to as the 'negative effects of climate variability and climate change that people have not been able to cope with or adapt to' (Warner et al., 2012, p. 20). These impacts may take the form of extreme weather events such as floods, droughts or hurricanes, as well as slow-onset events like sea-level rise, ocean acidification, and desertification. While academics and policy analysts have begun to pay more attention to the concept of loss and damage, especially since the Paris Agreement and its inclusion of Article 8 dedicated to the issue, the relationship between loss and damage and human rights remains significantly understudied. This is where this article seeks to contribute.

Different loss and damage scenarios impact human rights in various ways. These include impacts on *civil and political rights*, such as the right to life, liberty and property (Humphreys, 2009, p. 9), *economic, social and cultural rights* such as the right to work, education, social security, highest attainable standard of physical and mental health, adequate food, clothing and housing, and the continuous improvement of living conditions; as well as *collective rights*, including the right to development, self-determination, peace, a healthy environment, and minority rights more generally. While describing the concrete threats posed by each possible impact scenario goes beyond the scope of this article, previous studies have mapped the implications of climate impacts for the enjoyment of human rights (Martínez Blanco & Toussaint, 2018; UNEP, 2015; UNHRC, 2009a, paras. 21–41; 2018d). Ultimately, the threat posed by loss and damage to the fulfilment of a range of fundamental human rights, brings the issue within the scope of international human rights law, thus providing a clear normative basis for the adoption of an HRBA.

3. The state of policy integration

Having previously considered the human rights implications of loss and damage, the present section examines the state of policy integration of the international climate governance and human rights frameworks to highlight possible entry points for an HRBA to addressing loss and damage.

3.1. United Nations Framework Convention on Climate Change (UNFCCC)

The linkages between human rights and climate change did not suddenly emerge in the Paris Agreement; rather, they were the result of a long process of advocacy. The preamble of the United Nations Framework Convention on Climate Change (UNFCCC) states that climate change is a 'common concern for humankind', that 'may adversely affect natural ecosystems and humankind' (UNFCCC, 1992, preamble). Furthermore, Article 2 sets the time frame to reduce the causes of climate change with a view to ensuring 'that food production is not threatened and to enable economic development to proceed in a sustainable manner'. Though tacit, these references began to highlight the linkages between climate change and human wellbeing which provided fertile ground for the 'political narrative' (Mayer, 2016, p. 110) that led to the Paris Agreement's inclusion of human rights-related language.

The endeavour to have human rights explicitly included in treaty text dates back to the origins of the Convention, with the effort of the G77 to recognize the right to development in Article 2 as an 'inalienable human right', which was met with opposition from industrialized countries (Duyck, Jodoin, & Johl, 2018, p. 4). Mentions of human rights and rights-related language began to emerge in various forums, including at the Conferences of the Parties (COPs) to the UNFCCC (UNFCCC, 1995, Annex II, para. 11; 2008, paras. 121–122), in the Fourth Assessment Report of the Intergovernmental Panel on Climate Change (IPCC) with reference to human rights-related litigation (IPCC, 2007, p. 793), and in the resolutions and reports of the Human Rights Council (UNHRC, 2008), partly as a result of advocacy by non-governmental organizations (Rajamani, 2018, p. 239). However, it was not until Decision 1/CP.16 at COP 16 in Cancún in 2010 that the Parties adopted the first explicit reference to human rights regarding climate action. One possible reason why the recognition to fully respect human rights remained absent from the legal texts for such a long time was the 'disciplinary path-dependence' by practitioners – understanding human rights and climate change as two distinct fields – that took several years to overcome (Humphreys, 2009, p. 4).

3.2. International human rights bodies

Human rights and climate change governance remained the subject of different legal frameworks for many years (Savaresi, 2018) until a bridge was created through cross-pollination between the UN Human Rights Council (UNHRC) and the UNFCCC. This dynamic began in 2007 with the adoption by a group of Small Island Developing States (SIDS) of the Malé Declaration on the Human Dimension of Global Climate Change (SIDS, 2007). The declaration requested the UNFCCC to collaborate with the UNHRC and the UN Special Rapporteur Paul Hunt, who requested that the UNHRC study the impact of climate change on human rights generally and the right to health in particular (UNGA, 2007, para. 107(j)). This led to Resolution 7/23 (UNHRC, 2008), which asked the Office of the United Nations High Commissioner for Human Rights (OHCHR) to produce a report on the relationship between climate change and human rights (UNHRC, 2009a) and subsequently Resolution 10/4 welcoming the steps taken by the OHCHR and the UN Climate Change Secretariat to facilitate the exchange of information in the areas of human rights and climate change (UNHRC, 2009b). In 2010, the UNFCCC COP finally broke its silence on the issue by noting Resolution 10/4 in its preamble and emphasizing 'that Parties should, in all climate change-related actions, fully respect human rights' (Decision 1/CP.16, para. 8). The Paris Agreement's subsequent explicit reference to human rights is partly the result of this cross-pollination and of stakeholder advocacy and efforts such as the 'Geneva Pledge' (Human Rights and Climate Change Working Group, 2015) which lobbied to integrate human rights into the climate governance framework and created express linkages to both existing and future obligations of Parties regarding human rights (Duyck, Jodoin, et al., 2018, p. 5).

In recent years, international human rights bodies have begun to explore different ways human rights should be considered in the international policy response to loss and damage. For example, recent OHCHR reports have addressed the issue of human rights, climate change and persons displaced across international borders (OHCHR, 2016a), culminating in a UNHRC resolution on the issue (UNHRC, 2017b). The UNHRC further considered the rights of children in light of climate displacement (UNHRC, 2018a). Morever, former Special Rapporteur John Knox considered in his report on human rights obligations and the environment that the need to respond to climate change does not excuse a state from complying with its human rights obligations, and that it should take additional measures to protect the rights of those most vulnerable to environmental harm (UNHRC, 2018b, paras. 31, 40 and 54). In addition, the OHCHR recommended improving institutional arrangements and finance for the WIM, discussed below, considering that loss and damage is also concerned with the 'long-term impact [of climate change] on development, the ability of States to promote and protect human rights, and the availability of development assistance following losses related to climate change' (OHCHR, 2017, para. 58). Similarly, the UN Special Rapporteur on the Right to Food considered as essential the need for increased finance to support developing countries to deal with the long-term impacts of climate change on food security through adaptation and by addressing loss and damage (UNHRC, 2018c, para. 107).

3.3. Paris Agreement

With the Paris Agreement, human rights gained more space in climate governance by being mentioned in the treaty preamble with references to specific rights, including the right to development. Specifically, when implementing climate actions,

> Parties should [...] respect, promote and consider their respective obligations on human rights, the right to health, the rights of indigenous peoples, local communities, migrants, children, persons with disabilities and people in vulnerable situations and the right to development, as well as gender equality, empowerment of women and intergenerational equity. (Paris Agreement, preamble)

Moreover, Parties should be guided by the general objective of the UNFCCC of protecting food production and enabling sustainable development, and by the principle of equity, considering the benefit of present and future generations in accordance with Parties' common but differentiated responsibilities and respective capabilities (UNFCCC, 1992, Articles 2 and 3). The Paris Agreement also mandates the Parties to strengthen the response to climate change in the context of sustainable development (Article 2.1) and to reduce the risk of loss and damage as part of that mandate (Article 8.1).

Described by some authors as 'groundbreaking' (Savaresi, 2018, p. 35), these gains are relative as the Agreement does not provide an explicit way to operationalize or monitor human rights. Essentially a compromise outcome (Savaresi & Hartmann, 2015, p. 2), the inclusion of human rights in the preamble, and human rights-related language in the operative text, makes the issue of human rights more visible, filling the historical gap left by the UNFCCC and Kyoto Protocol (Harmeling, 2018, p. 93). This is an advantage for stakeholders engaged in ongoing advocacy campaigns such as the 'Great 8' (CAN, 2018) that strive to include human rights in the planning of Parties' Nationally Determined Contributions (NDCs) and seek ways to operationalize and monitor human rights in climate action. The 'Great 8' comprise fundamental elements intended to put people at the centre of climate policies and actions. These elements include 'human rights, indigenous peoples' rights, public participation, gender, just transition, food security, ecosystem integrity and the protection of biodiversity, and intergenerational equity' (Human Rights and Climate Change Working Group, 2018).

The cross-pollination between the climate regime and other policy frameworks will continue to play an important role in strengthening the protection of human rights in climate action. As Savaresi (2018, p. 38) notes, the exchange between institutions, publics and legal frameworks can 'be systematised and become instrumental to the streamlining of human rights considerations into the climate regime'. For example, the inclusion of human rights in the Sustainable Development Goals (SDGs) (UNGA, 2015) further strengthens the arguments for a human rights-based approach to climate action under the Paris Agreement. Goal 13 on climate action, in particular, acknowledges the UNFCCC as the primary forum to address climate change (para. 31) and reciprocally the COP welcomed the SDGs in the Paris COP decision (Decision 1/CP.21, preambular recital 4). Moreover, the promotion of sustainable development is one of the principles guiding Parties in implementing the UNFCCC (UNFCCC, 1992, Article 3.4), and is considered within the objective of the Paris Agreement (Article 2.1). Relying on the international law principle of systemic integration, further discussed below, it follows that the references to sustainable development in the operative text of the Paris Agreement must be read in light of the SDGs and their respective references to human rights (Ferreira, 2016, p. 10). This signifies that policymakers should integrate a human rights-based approach to satisfy the sustainability requirements of the Paris Agreement.

The Paris Agreement Work Programme (known informally as the 'Paris Rulebook'), which was adopted at COP 24 in 2018, provided an important opportunity to flesh out the substantive mandate of Article 8, anchor loss and damage in the treaty's mechanisms, and adopt a human rights perspective as suggested by the former UN Special Rapporteur John Knox (UNHRC, 2016, para. 64). Going into COP 24, several Parties and observers demanded that the Rulebook give effect to the language around human rights and the rights-related provisions contained in the preamble of the Agreement. Despite advocacy efforts before and during the Katowice negotiations, proposals to include an explicit reference to *human rights* in the Rulebook ultimately remained unsuccessful. Instead, it vaguely refers to 'other contextual aspirations and priorities acknowledged when joining the Paris Agreement' in relation to NDC guidance (Decision 4/CMA.1, Annex I, para. 4(a)(ii)(c)) and in other instances includes references to specific human rights-related elements such as gender, indigenous peoples, and stakeholder involvement. A full analysis of the human rights-related provisions of the Paris Rulebook goes beyond the scope of this article and has been provided by others (CIEL, 2018). The inclusion of at least some human rights-related language in the Rulebook could be considered a positive development as it serves as a reminder to public and private actors to consider human rights when implementing climate actions, including loss and damage policies and actions. At the same time, while most of the references are weakly formulated and part of a non-binding package of COP decisions, Parties are legally bound by their existing obligations under international and regional human rights treaties, and should be guided by the Paris Agreement's preamble.

3.4. Warsaw International Mechanism

The WIM is the central body mandated under the UNFCCC and Paris Agreement to deal with loss and damage associated with climate change impacts and is thus a key process for protecting the human rights of affected communities. However, there is at present little explicit human rights language in COP decisions on loss and damage and in the technical work of the WIM. Harmeling (2018) suggests that the lack of explicit human rights language on loss and damage in COP decisions is not surprising given the limited engagement with human rights concepts in the UNFCCC. However, while neither of the consecutive workplans of the WIM's Executive Committee (Excom) contains explicit human rights language, its scope of work is *de facto* concerned with the human rights implications of loss and damage. For example, Action Area 1 of the initial two-year workplan requires the Excom to work on enhancing understanding of how loss and damage affects 'particularly vulnerable developing countries, segments of the population that are already vulnerable owing to geography, socioeconomic status, livelihoods, gender, age, indigenous or minority status or disability, and the ecosystems that they depend on' (UNFCCC, 2014, p. 195). With explicit rights language few and far between in the WIM's outcomes, there is scope to streamline human rights into the work of the Excom and its implementing bodies.

Significant strides have been made in terms of policy integration under the WIM's Task Force on Displacement (TFD). This could provide a useful model to follow. At COP 24, Parties welcomed the Excom's annual report that included a set of recommendations on the displacement of people due to climate change impacts, based on the work of the TFD. The recommendations invite Parties to formulate laws and policies on displacement, 'taking into consideration their respective human rights obligations and, as appropriate, other relevant international standards and legal considerations' (Decision 10/CP.24, Annex, para. 1(g)(i)). Other instances where human rights are mentioned expressly refer to the emerging work of other institutions, most prominently the OHCHR. For example, the OHCHR has been vocal in trying to integrate its work on human rights into the work of the TFD. Several Parties, experts, and legal scholars have called for greater integration of human rights in the UNFCCC's work on displacement (Atapattu, 2009; IOM, 2017; Limon, 2009; Lyster, 2015; Mayer, 2012; UNFCCC, 2017a).

Ultimately, the enhancement and strengthening of the WIM in accordance with Article 8.2 of the Paris Agreement, and the further operationalization of its mandate to enhance action and support, including finance, technology and capacity-building (Decision 3/CP.22, para. 4) will be essential to protect human rights in the loss and damage context. Recognizing this, several civil society organizations called in their submissions to the WIM for a human rights-based and gender equitable approach to loss and damage finance (UNFCCC, 2017b).

4. Rationale for adopting a human rights-based approach to loss and damage

4.1. Origins and definition

In 2003, several UN bodies jointly developed the concept of an HRBA in an attempt to streamline engagement with human rights across UN agencies in the field of international development (OHCHR, 2006). Specifically, these UN bodies outlined six guiding principles that should be considered in development policies: (1) universality and inalienability, (2) indivisibility and (3) inter-dependence and inter-relatedness of human rights; (4) equality and non-discrimination; (5) participation and inclusion; and (6) accountability and the rule of law (HRBA Portal, n.d.). On the back of this work, the OHCHR defines an HRBA as a 'conceptual framework for the process of human development that is normatively based on international human rights standards and operationally directed to promoting and protecting human rights' (OHCHR, 2006, p. 15). Broberg and Sano (2018, p. 667) further suggest that an HRBA creates a 'methodological framework for the realization of human rights' by 'linking the normative basis to its concrete implementation'. However, they note that in practice, HRBAs may differ significantly as they are determined by the variety of actors involved and other contextual factors.

HRBAs are already being advocated as strategic tools in the context of international climate policy, particularly in relation to climate mitigation and adaptation. Duyck, Jodoin et al. (2018, p. 7) suggest that human rights are frequently championed in the climate policy context to 'ensure that laws, policies, programmes, and projects adopted to mitigate or adapt to climate change respect, protect, and fulfil human rights'. Rajamani (2010, p. 395) relevantly points to the strategic nature of the approach, which

> seeks a reframing of the climate problem that would provide nations with a 'compass for policy orientation' and draw them towards ever more stringent actions. The focus […] is on the broader range of human rights placed at risk by the impacts of climate change and the ethical pull this might create.

As will be explored below, an HRBA goes beyond a mere mainstreaming of human rights into climate negotiation outcomes and the wider climate policy discourse.

4.2. International law context

Crucially, an HRBA emphasizes the obligations states already have under existing international and regional human rights treaties. It thus serves as a reminder that climate policy does not exist in a vacuum, in isolation from general obligations of international law (Burkett, 2009), in particular, human rights law. While subject to limitations, international human rights law has the potential to provide judicial recourse in the absence of a liability and compensation regime for loss and damage under the climate change regime. By highlighting the existing linkages between loss and damage and human rights obligations in international and regional human rights instruments, the HRBA provides an essential tool to affected communities and civil society to shape public opinion on climate policy. Human rights bodies and courts can play a crucial role in ensuring accountability and enforcement of affected human rights (De Schutter, 2012), and could thus provide a complementary tool to the HRBA (see also Pekkarinen, Toussaint, & van Asselt, 2019). In this sense, adopting an HRBA should be understood as a preventive measure, targeting the design, implementation, and evaluation of loss and damage policies and actions, rather than addressing human rights violations that may arise from loss and damage.

An HRBA rests on the international law principle of systemic integration, which provides an important tool for interpreting the human rights-related provisions in the UNFCCC and Paris Agreement in the wider context of human rights law. The principle is reflected both in the theory and practice of international law (Sheeran,

2014). It is laid down in Article 31(3)(c) of the 1969 Vienna Convention on the Law of Treaties (VCLT), which states that the interpretation of treaties and their mandate cannot be read in isolation but shall take into account 'any relevant rules of international law applicable in the relations between the parties', thus including international human rights law. The climate change regime cannot be understood as being limited to its objectives because:

> law is also about protecting rights and enforcing obligations, above all rights and obligations that have a backing in something like a general, public interest. Without the principle of *'systemic integration'* it would be impossible to give expression to and to keep alive, any sense of the common good of humankind, not reducible to the good of any particular institution or 'regime'. (ILC, 2006, para. 480, authors' emphasis)

This argument is further bolstered by the fact that in accordance with Article 31(2) VCLT, the operative text of the Paris Agreement – and by extension also the Paris Rulebook – needs to be interpreted in light of the preamble, including the preambular references to human rights and to other rights-related provisions. Based on the principle of systemic integration, this requires a consideration of relevant international and regional human rights instruments (see also Duyck, 2017, pp. 11–12). This was affirmed in the recent Advisory Opinion by the Inter-American Court of Human Rights (IACHR) which took an expansive view on the body of international human rights law based on the principle of systemic integration (IACHR, 2017, para. 45).

International human rights law thus provides an important normative framework for states in implementing climate policies and actions (UNHRC, 2009a, para. 71). Aside from individual entitlements, this normative framework provides clear guidance on the human rights obligations of Parties under relevant international and regional human rights treaties they have ratified. Of particular relevance to the loss and damage context is the duty of states under international human rights law to assist and cooperate in the protection of those rights (UNGA, 1970; UN Charter, 1945, Articles 1.3, 55(c) and 56). This duty extends to facilitating the fulfilment of human rights in other countries (UNHRC, 2009a, para. 86). The implementation of an HRBA under the climate change regime aims to analyse 'obligations, inequalities and vulnerabilities, and seek[s] to redress discriminatory practices and unjust distributions of power. It [would] anchor[s] plans, policies, and programmes in a system of rights, and corresponding obligations established by international law' (OHCHR, 2016b, p. 4). Adopting an HRBA could thus serve to anchor this normative framework enshrined in human rights law in the international community's policy response to loss and damage.

4.3. Relevance for loss and damage

Whilst the adoption of an HRBA has been advocated in relation to climate mitigation and adaptation (CIEL, 2009; Knur, 2014; Limon, 2009), arguably, its particular value lies in calling attention to the human rights impacts of climate change manifesting themselves in the form of loss and damage. Both mitigation and adaptation measures are essential to reducing the threat posed by climate impacts to human rights and could themselves adversely affect these rights. Loss and damage disproportionately affects poor and marginalized persons (Cameron, Shine, & Bevins, 2013), and adopting an HRBA specifically on loss and damage focuses attention on these persons. As Broberg and Sano (2018, p. 675) note, an HRBA 'gives highly relevant experience in how legal instruments can be used in relation to poverty and marginalisation'.

Applied to the context of loss and damage under the UNFCCC, an HRBA could help refocus the political narrative on the fundamental rights of the individual that must be protected through, and when taking, climate actions in accordance with a pre-established normative framework. In this sense, adopting an HRBA could remedy a central deficiency of the ongoing policy debate, which frames loss and damage in abstract, state-centric terms as a developing country issue (Toussaint, 2019). By calling attention to the fundamental human rights of the individual, the approach requires consideration of the intersectionality of loss and damage impacts, including questions of race, gender, class, age, and economic well-being. As noted by former UN Special Rapporteur John Knox, States have heightened duties with respect to members of vulnerable groups and 'should seek to protect the most vulnerable in developing and implementing all climate-related actions' (UNHRC, 2016, paras. 81–82). In this light, an HRBA on loss and damage has the potential to both shape climate policies and hold countries accountable for their climate commitments (UNHRC, 2017a, para. 31).

In particular, the HRBA principle of participation and inclusion – which derives its normative basis from relevant international and regional legal instruments[1] – provides valuable guidance to policymakers and implementers of loss and damage actions. Applied to the loss and damage context, it requires states to ensure the effective participation of those most directly affected in the design and implementation of loss and damage policies and actions. The HRBA could thus be seen as a democratizing tool in the field of climate governance given that the voices of poor and marginalized communities most affected by loss and damage have thus far been largely absent from the UNFCCC (Toussaint, 2019). This reflects a recognition that actual and potential victims of loss and damage should be empowered as 'active participants' (Broberg & Sano, 2018, p. 670) in decisions that concern their lives and livelihoods. By going beyond mere rhetoric, an HRBA promotes the institutionalization of cooperation between states and those most affected (Broberg & Sano, 2018, p. 669) at the international level. The HRBA could thus prove a strategic tool to significantly strengthen the international response to loss and damage in terms of human rights and to remind policymakers and implementers that international climate policy does not operate in a vacuum.

4.4. Limitations

However, an HRBA is no silver bullet and is subject to limitations. One drawback of the approach lies in its evidently anthropocentric and narrow temporal focus, that is, its failure to engage with broader conceptions of the rights of nature, other non-human-centred approaches or the interests of future generations. As with any rights-based approach, the adoption of an HRBA does not automatically guarantee the enforcement of those rights. In practice, and depending on the rule of law, some duty-bearers (e.g. national or local governments) against whom these rights are usually enforced, may not see their enforcement as a top priority or may lack the capacity to enforce them (Broberg & Sano, 2018). Enforcement may be even more difficult at the international level for affected individuals seeking to hold states to account before international courts and tribunals (Pekkarinen et al., 2019). As Hickey and Mitlin (2009) observe in the development context, the adoption of an HRBA risks promoting inequalities, conflicts and the unsustainable use of natural resources. Also, an HRBA does not provide detailed guidance for planning policies and actions on the ground. In fact, many civil society, grassroot, and community-based organizations have been cautious in their human rights advocacy in order not to upset their relationships with local and national governments (Christoplos, Funder, McGinn, & Wairimu, 2014, p. 3).

5. Adopting a human rights-based approach to loss and damage under the climate change regime

In practice, adopting an HRBA to loss and damage under the climate change regime would involve monitoring and evaluating climate actions and their outcomes based on human rights standards and principles; recognizing affected persons (rights-holders) as active participants in climate decision-making; assessing the capacity of rights-holders to claim their rights and the capacity of governments (duty-bearers) to fulfil their human rights obligations; providing capacity-building and learning opportunities, transparency of process and access to information; and ensuring climate actions are informed by the recommendations of relevant human rights bodies (based on HRBA Portal, n.d.; OHCHR, 2010, n.d.). We argue that this approach should be based on human rights law and follow the six guiding principles outlined by the UN.

Specifically, the approach could take the following form: The inclusion of human rights impact assessments (Walker, 2005, p. 217) for loss and damage policies and actions; development of human rights guidelines applicable to loss and damage governance and actions, including monitoring and reporting (Duyck, Lennon, Obergassel, & Savaresi, 2018, p. 198); periodic reviews by human rights bodies of broader loss and damage governance (e.g. the WIM) as well as of loss and damage policies and actions implemented under the Paris Agreement; and the creation of a procedure to monitor compliance. This list is neither intended to be prescriptive nor exhaustive but represents an attempt to provide practice examples that could make a material contribution to strengthening the UNFCCC policy response to loss and damage. In the following, we identify possible entry points for an international HRBA to loss and damage.

5.1. The Warsaw International Mechanism

There is significant scope to enhance the integration of an HRBA in the loss and damage work of the WIM to protect human rights in the loss and damage context. The necessary mandate is already in place. Article 8.5 of the Paris Agreement requires the WIM to collaborate with existing bodies and expert groups under the Agreement, as well as relevant organizations and expert bodies outside the Agreement. COP Decision 2/CP.19 further stipulates that the WIM should complement and draw upon the work of these bodies, and involve them in its work (para. 6).

By the same token, the COP invites relevant international and regional organizations, institutions and processes to integrate measures to address loss and damage in their work and strengthen synergies (Decision 2/CP.19, para. 11). The work done by the OHCHR, particularly in the context of climate displacement could serve as an example to follow. The WIM could play an important role in promoting policy coherence (Decision 2/CP.19, para. 12) by 'fostering dialogue, coordination, coherence and synergies among all relevant stakeholders, institutions, bodies, processes, and initiatives outside the Convention, with a view to promoting cooperation and collaboration' (Decision 2/CP.19, para. 5(b)(ii)).

One possible avenue could be for the COP to request the WIM Excom to develop guidelines in the form of a recommendation to be adopted by the COP. Specifically, this could include guidelines for adopting an HRBA to loss and damage policies and actions, including monitoring and reporting thereof, as well as guidelines for conducting human rights impact assessments of loss and damage projects prior to implementation. On the one hand, these guidelines could serve the WIM's mandate of enhancing knowledge and understanding by addressing gaps in understanding and expertise and providing best practices (Decision 2/CP.19, para. 5(a)(i) and (iii)). Moreover, the guidelines could help to strengthen dialogue, coordination, coherence, and synergies among relevant stakeholders by providing oversight under the UNFCCC on the assessment and implementation of approaches to address loss and damage and by engaging with processes outside the Convention (Decision 2/CP.19, para. 5(b)(i-ii)). On the other hand, the guidelines could serve the WIM's mandate to enhance action and support by providing technical guidance and securing expertise to facilitate the development of additional approaches to address loss and damage (Decision 2/CP.19, para. 5(c)(i-iii)) – in this case an HRBA. Furthermore, Parties and observers should be encouraged to call on the Excom and the Technical Expert Group on Comprehensive Risk Management Approaches to integrate human rights impact assessments into the work under Strategic workstream (c) of the WIM's five-year workplan on enhanced cooperation and facilitation in relation to comprehensive risk management approaches.

Another, more ambitious, measure could be for the Excom to create a specialized body (e.g. a task force or advisory group) on human rights in the context of loss and damage (in line with Decision 2/CP.20, para 8). This body could be composed of relevant experts and stakeholders with the active participation of persons affected by loss and damage and could support the work of the Excom on furthering an HRBA by preparing the draft guidelines on loss and damage mentioned above. Moreover, the specialized body could be tasked with monitoring Parties' compliance with the proposed guidelines within the limitations of the WIM's mandate.

5.2. The Paris Agreement and its rulebook

With its Rulebook mostly adopted,[2] the onus is now on Parties to push ahead with the implementation of the Paris Agreement. Considering the treaty's preambular reference to human rights and the fact that Parties are bound by their existing human rights obligations, the Paris Agreement and its Rulebook can be a powerful tool for promoting HRBA in the treaty's implementation and for strengthening the integration of human rights in loss and damage policy.

5.2.1. Nationally determined contributions

The NDCs represent a cornerstone of the Paris Agreement architecture. While the Paris Rulebook does not specify loss and damage or human rights as a component of Parties' NDCs, it does not prohibit them from including such elements either (Decision 4/CMA.1, para. 8). In fact, 49 countries already refer to loss and damage in their NDCs, the majority being least developed countries (LDCs) and SIDS, and none of them developed countries (Kreienkamp & Vanhala, 2017; UNFCCC, 2018). Not only do LDCs and SIDS experience loss and damage most severely today, they also tend to lack adequate resources to deal with these impacts without international support. However, given the projected scale and intensity of loss and damage resulting from temperature rise beyond 1.5°C, loss and damage is not only an issue of relevance for developing countries (Toussaint, 2019). All Parties potentially affected, including developed countries, could opt to report information on averting, minimizing and addressing loss and damage in their NDCs, including, where relevant, support needed and provided.

Likewise, there is further scope for integrating human rights reporting into NDCs. Only 17 Parties have thus far included human rights as a guiding principle for implementing their NDCs, while seven Parties cite human rights as relevant to their domestic context (Duyck, Lennon, et al., 2018). Beyond merely mentioning human rights, Parties could better integrate an HRBA by including human rights impact assessments in their NDCs. This would lay down important criteria to shape their national policies and actions on loss and damage to conform with an HRBA. Continued advocacy by human rights groups at the international level and especially domestically by civil society will be crucial as countries revise their NDCs.

5.2.2. Transparency framework

The Enhanced Transparency Framework for Action and Support is mandated to 'provide a clear understanding of climate change action in the light of the objective of the Convention as set out in its Article 2' (Paris Agreement, Article 13.5). The Transparency Framework guidelines adopted in the Paris Rulebook include loss and damage under Article 7 on adaptation (Decision 18/CMA.1, Annex, para. 115).[3] This is perhaps unsurprising since Article 13.8 of the Paris Agreement asked Parties to provide 'information related to climate change impacts', which arguably encompasses loss and damage, under Article 7. Although there was no express mandate to include loss and damage in the Transparency Framework, the issue thus effectively piggybacked its way into the Paris Rulebook on the back of adaptation. Therefore, notwithstanding the lack of political will to formalize the inclusion of loss and damage in the Transparency Framework by expressly referring to Article 8, the necessity of reporting information related to loss and damage prevailed.

Regardless of whether it is included under Article 7 or Article 8, the inclusion of loss and damage in the Transparency Framework is a positive development because it reflects a recognition that information on loss and damage is crucial to the 'effective implementation' of the Paris Agreement in line with Article 13.1. Furthermore, the Transparency Framework should include information about the NDCs and adaptation to inform the Global Stocktake (GST) (Article 13.5). Parties can also include loss and damage as additional information in the facilitative, multilateral consideration of progress under the Transparency Framework (Decision 18/CMA.1, Annex, Section VIII.B., para. 190(c)). The inclusion of human rights impact assessments in NDCs and the development of human rights guidelines for reporting and monitoring loss and damage policies and actions would serve to generate specialized information that could be fed into the Transparency Framework.

5.2.3. Global stocktake

The Paris Rulebook stipulates that the GST will take into account, 'as appropriate', efforts to avert, minimize and address loss and damage through its technical dialogue feeding into a technical assessment (Decision 19/CMA.1, para. 6(b)(ii)). Furthermore, the GST will consider information at a collective level related to '[e]fforts to enhance understanding, action and support, on a cooperative and facilitative basis, related to averting, minimizing and addressing loss and damage' (Decision 19/CMA.1, para. 36(e)). The Rulebook further invites constituted bodies – including the WIM – to prepare a synthesis report in their area of expertise (Decision 19/CMA.1, para. 24) which could include information based on the human rights guidelines proposed in this article.

This represents an opportunity to complement the information on loss and damage provided to the Transparency Framework under Article 7 and the NDCs. If countries adopt an HRBA to loss and damage and report on it in their NDCs or other forms of input into the Transparency Framework, then this information can be submitted to the technical assessment and technical dialogue that informs the GST. Together with the Transparency Framework, the GST can thus serve to better understand and monitor loss and damage actions undertaken, or the lack of them, having regard to the human rights of people that experience loss and damage due to the adverse effects of climate change.

5.3. UN climate change secretariat

During a side event at COP 24 in 2018, Parties and stakeholders repeated their calls for the creation of a Human Rights Focal Point at the UN Climate Change Secretariat (Mary Robinson Foundation – Climate Justice, 2018). Such a focal point could be modelled after the UNFCCC Gender Focal Point and could build capacity on human rights in the Secretariat, help Parties to plan and implement human rights-based climate actions, and liaise with OHCHR and other relevant bodies on human rights and climate change. Loss and damage could be introduced by the Human Rights Focal Point as an element in policy discussions among human rights institutions on other key issues, such as women's rights, the rights of indigenous peoples, and cultural rights (in particular in the context of non-economic loss and damage), as well as gender-responsive approaches in climate action which, in turn, inform the normative basis for the HRBA.

5.4. International human rights bodies

International human rights institutions will play a crucial role in implementing an HRBA to loss and damage under the climate change regime. In practice, the degree to which states will support the implementation of an HRBA will vary, depending on their national priorities, democratic configurations, and the enforcement of relevant international and regional human rights treaties they have ratified. Knur (2014, p. 56) observes that '[h]ow States use their margin of appreciation requires monitoring, and thus the relationship of human rights and climate change needs to be (re-) discovered by the pertinent human rights institutions'. The monitoring function can be fulfilled by the state reporting procedure under the UNHRC and the convention committees of major international human rights treaties, which are composed of international experts. On the one hand, the convention committees have the competence to evaluate how climate change affects the fulfilment of human rights under the relevant convention (Knur, 2014, p. 45), which could include a more specific assessment of loss and damage. On the other hand, states could be required to report on their progress in fulfilling their duty of international cooperation and assistance, specifically on how their efforts to address loss and damage under the WIM and mechanisms of the Paris Agreement are human rights-compliant (building on Knur, 2014, p. 50). Furthermore, human rights institutions have the competence to prepare *general comments* which clarify the scope of convention rights and provide policy recommendations (Knur, 2014, p. 46). While not legally binding, this is a readily available tool to strengthen the integration of an HRBA by concretizing state obligations in relation to human rights law on climate loss and damage. Lastly, international human rights bodies could cooperate with the Human Rights Focal Point, proposed above, to prepare periodic human rights reviews of loss and damage policy and actions under the climate change regime to further inform the implementation of an HRBA to loss and damage and complement the work of the WIM.

6. Conclusion

This article proposed the adoption of an HRBA to address loss and damage under the climate change regime. It explored the human rights implications of loss and damage and provided an assessment of the state of policy integration of human rights and climate governance in this context. The article found that adopting an HRBA can be an important strategic tool for policymakers to strengthen the international climate change response on loss and damage. Such an approach builds on the existing human rights obligations of Parties under international law and should be based on the six guiding principles set out by the UN. The article thus highlighted opportunities for anchoring an HRBA under the WIM and key mechanisms for the implementation of the Paris Agreement, specifically the NDCs, Transparency Framework and GST, as well as for creating a Human Rights Focal Point at the UN Climate Change Secretariat. Conversely, the article also considered the integration of loss and damage in the work of relevant human rights bodies, specifically the UNHRC and the OHCHR and their contribution to strengthening the normative basis of an HRBA to loss and damage under the climate change regime.

Considering also the outcomes of COP 24, the article discussed the relevance of the Paris Rulebook for advancing the work on loss and damage and human rights. It found that even in the absence of explicit human rights references in the Rulebook, Parties are bound by their existing obligations under the international and regional human rights treaties they have ratified and should be guided by the Paris Agreement's preamble in the implementation of climate actions. While the Rulebook does not include a reference to Article 8 of the Paris Agreement, loss and damage is still included in the Transparency Framework (under the adaptation section) and the GST. Furthermore, despite the political disagreements that left Article 8 sidelined and human rights explicitly out of the Rulebook, advocacy on loss and damage and human rights must continue and expand on the foundations the Paris Agreement has provided. After all, international cooperation on loss and damage is an essential recourse for those most vulnerable to climate change, and an HRBA sets the standard that Parties should live up to.

Notes

1. Including the 1992 Rio Declaration on Environment and Development, Principle 10; the UNECE Convention on Access to Information, Public Participation in Decision-making and Access to Justice in Environmental Matters, Aarhus 25 June 1998 (Aarhus Convention, in force 30 October 2001); the UNECLAC Regional Agreement on Access to Information, Public Participation and Justice in Environmental Matters in Latin America and the Caribbean, Escazú, 4 March 2018 (Escazú Agreement, not yet in force); and the Almaty Guidelines on Promoting the Application of the Principles of the Aarhus Convention in International Forums, UN Doc. ECE/MP.PP/2005/2/Add.5, 20 June 2005, Almaty 27 May 2005, Annex.
2. With the exception of modalities on Article 6 (cooperative mechanisms) which were deferred until 2019 and guidance on Article 4 (NDC features) deferred until 2024 and Article 4.10 (common timeframes for NDCs) to be streamlined from 2031.
3. An earlier draft of the Rulebook proposed including loss and damage in the Transparency Framework under adaptation, financial instruments, technology transfer and capacity-building support (UN Climate Change Secretariat, Additional tool under item 5 of the agenda, APA1.6.Informal.1.Add.3 (3 August 2018)). Retrieved from https://unfccc.int/sites/default/files/resource/APA1.6.Informal.1.Add_.3.pdf.

Acknowledgments

The authors would like to express their gratitude to the Alexander von Humboldt Foundation and the Institute for Advanced Sustainability Studies. We further thank the editors of this special issue and the four anonymous referees for their helpful guidance. Many thanks are also due to Harro van Asselt for his useful comments and to Cecilia Oliveira for her comments on an earlier draft of the paper.

Disclosure statement

No potential conflict of interest was reported by the authors.

ORCID

Patrick Toussaint (iD) http://orcid.org/0000-0003-1314-4946
Adrian Martínez Blanco (iD) http://orcid.org/0000-0002-6787-8366

References

Atapattu, S. (2009). Climate change, human rights, and forced migration: Implications for international law. *Wis. Int'l LJ*, 27, 607–636.
Broberg, M., & Sano, H.-O. (2018). Strengths and weaknesses in a human rights-based approach to international development – an analysis of a rights-based approach to development assistance based on practical experiences. *The International Journal of Human Rights*, 22(5), 664–680. doi:10.1080/13642987.2017.1408591
Burkett, M. (2009). Climate reparations. *Melbourne Journal of International Law*, 10, 509–542.
Cameron, E., Shine, T., & Bevins, W. (2013). *Climate justice: Equity and justice informing a new climate agreement*. Retrieved from http://www.wri.org/sites/default/files/climate_justice_equity_and_justice_informing_a_new_climate_agreement.pdf
CAN. (2018). The "Great 8" for people-centered climate action. *CAN International*. Retrieved from http://eco.climatenetwork.org/sb48-eco5-7/
Christoplos, I., Funder, M., McGinn, C., & Wairimu, W. (2014). The role of human rights in climate change adaptation: evidence from civil society in Cambodia and Kenya.
CIEL. (2009). *Practical approaches to integrating human rights and climate change law and policy*. Geneva. Retrieved from https://www.ciel.org/reports/practical-approaches-to-integrating-human-rights-and-climate-change-law-and-policy-february-2009-orellana-3/
CIEL. (2018). *Report from the Katowice climate conference: Promoting human rights in climate action at COP-24*. Geneva. Retrieved from https://www.ciel.org/wp-content/uploads/2018/12/HRCC_Roundletter_COP24_December2018.pdf

De Schutter, O. (2012, April 24). Climate change is a human rights issue – and that's how we can solve it. *The Guardian*. Retrieved from www.theguardian.com/environment/2012/apr/24/climate-change-human-rights-issue

Duyck, S. (2017). The Paris climate agreement and the protection of human rights in a changing climate. *Yearbook of International Environmental Law, 26*, 3–45. Retrieved from https://dx.doi.org/10.1093/yiel/yvx011

Duyck, S., Jodoin, S., & Johl, A. (2018). Integrating human rights in global climate governance: An introduction. In S. Duyck, S. Jodoin, & A. Johl (Eds.), *Handbook of Human Rights and Climate Governance* (pp. 3–15). London: Routledge.

Duyck, S., Lennon, E., Obergassel, W., & Savaresi, A. (2018). Human Rights and the Paris Agreement's implementation Guidelines: Opportunities to develop a Rights-based approach. *Carbon and Climate Law Review, 12*(3), 191–202. doi:10.21552/cclr/2018/3/5

Ferreira, P. G. (2016). *Did the Paris agreement fail to incorporate human rights in operative provisions? Not if you consider the 2016 SDGs.* Retrieved from https://www.cigionline.org/sites/default/files/documents/Paper%20no.113.pdf

Harmeling, S. (2018). Climate change impacts: Human rights in climate adaptation and loss and damage. In *Handbook of Human Rights and climate Governance* (pp. 90–109). Llondon: Routledge.

Hickey, S., & Mitlin, D. (2009). The potential and pitfalls of rights-based approaches to development. In S. Hickey & D. Mitlin (Eds.), *Rights-based approaches to development: Exploring the potential and pitfalls* (pp. 209–230). Sterling, VA: Kumarian Press.

HRBA Portal. (n.d.). The human rights based approach to development cooperation: Towards a common understanding among UN agencies. Retrieved from http://hrbaportal.org/the-human-rights-based-approach-to-development-cooperation-towards-a-common-understanding-among-un-agencies

Human Rights and Climate Change Working Group. (2015). Promoting the Geneva pledge for human rights in climate action – human rights & climate change. Retrieved from http://climaterights.org/our-work/unfccc/geneva-pledge/

Human Rights and Climate Change Working Group. (2018). The great eight: calling for human rights principles in the Paris Agreement's Work Programme.

Humphreys, S. (2009). Introduction: Human rights and climate change. In S. Humphreys (Ed.), *Human Rights and climate change* (pp. 1–34). Cambridge: Cambridge University Press.

ILC. (2006). *Fragmentation of international law: Difficulties arising from the diversification and expansion of international law.* A/CN.4/L.682. Retrieved from http://legal.un.org/ilc/documentation/english/a_cn4_l682.pdf

Inter-American Court of Human Rights. (2017). *Advisory opinion OC-23/17 requested by the Republic of Colombia, 'Environment and Human Rights'.*

IOM. (2017). *Submission from the International Organization for Migration (IOM)*. Retrieved from https://unfccc.int/files/adaptation/groups_committees/loss_and_damage_executive_committee/application/pdf/iom_submission.pdf

IPCC. (2007). Climate change 2007: Mitigation. Contribution of working group III to the fourth assessment report of the Intergovernmental panel on climate change climate change 2007: Mitigation. In *Contribution of working group III to the fourth assessment report of the intergovernmental panel on climate change* (pp. 793–794). Cambridge: Cambridge University Press.

IPCC. (2018). Sustainable development, Poverty Eradication and Reducing Inequalities *Global Warming of 1.5°C*. In *An IPCC special report on the impacts of global warming of 1.5°C above pre-industrial levels and related global greenhouse gas emission pathways, in the context of strengthening the global response to the threat of climate change, sustainable development, and efforts to eradicate poverty* (pp. 454–456). Cambridge: Cambridge University Press.

Knur, F. (2014). The United Nations Human Rights-Based approach to climate change – Introducing a human dimension to international climate law. In S. von Schorlemer & S. Maus (Eds.), *Climate change as a Threat to Peace* (pp. 37–60). Bern: Peter Lang GmbH.

Kreienkamp, J., & Vanhala, L. (2017). *Climate change loss and damage*. Retrieved from https://www.ucl.ac.uk/global-governance/sites/global-governance/files/policy-brief-loss-and-damage.pdf

Lees, E. (2017). Responsibility and liability for climate loss and damage after Paris. *Climate Policy, 17*(1), 59–70.

Limon, M. (2009). Human rights and climate change: Constructing a case for political action. *Harv. Envtl. L. Rev, 33*, 439–476.

Lyster, R. (2015). Protecting the human rights of climate displaced persons: The promise and limits of the United Nations framework convention on climate change. In A. K. Grear, & J. Louis (Eds.), *Research Handbook on Human Rights and the Environment* (pp. 423–448). UK: Edward Elgar Publishing.

Martínez Blanco, A., & Toussaint, P. A. (2018). *TALANOA input on loss and damage and human rights*. Retrieved from https://unfccc.int/documents/184126

Mary Robinson Foundation – Climate Justice. (2018, 7 December). Building capacity for integrating human rights into climate action. Retrieved from https://www.mrfcj.org/resources/building-capacity-for-integrating-human-rights-into-climate-action-2/

Mayer, B. (2012). Sustainable development law on environmental migration: The story of an obelisk, a bag of marbles, and a tapestry. *Environmental Law Review, 14*(2), 111–133.

Mayer, B. (2016). Human rights in the Paris Agreement. *Climate Law, 6*(1-2), 109–117.

Mechler, R., Calliari, E., Bouwer, L. M., Schinko, T., Surminski, S., Linnerooth-Bayer, J., … Deckard, N. D. (2019). Science for loss and damage. In *Findings and propositions loss and damage from climate change* (pp. 3–37). Cham: Springer.

OHCHR. (2006). *Frequently asked questions on a human rights-based approach to development cooperation*. Retrieved from http://www.ohchr.org/Documents/Publications/FAQen.pdf

OHCHR. (2010). *Applying a human rights-based approach to climate change negotiations, policies and measures*. Retrieved from https://hrbaportal.org/wp-content/files/InfoNoteHRBA1.pdf

OHCHR. (2016a). *Analytical study on the relationship between climate change and the human right of everyone to the enjoyment of the highest attainable standard of physical and mental health.* A/HRC/32/23. Retrieved from https://undocs.org/A/HRC/32/23

OHCHR. (2016b). *Taking action on human rights and climate change. [Discussion Paper].* Retrieved from https://www.ohchr.org/Documents/Issues/ClimateChange/EM2016/TakingAction.pdf

OHCHR. (2017). *Summary of the panel discussion on human rights, climate change, migrants and persons displaced across international borders.* A/HRC/37/35. Retrieved from https://environmentalmigration.iom.int/sites/default/files/A_HRC_37_35_AEV.PDF

OHCHR. (n.d.). Human rights and climate change. Retrieved from https://www.ohchr.org/EN/Issues/HRAndClimateChange/Pages/HRClimateChangeIndex.aspx

Pekkarinen, V., Toussaint, P., & van Asselt, H. (2019). Loss and damage after Paris: Moving beyond rhetoric. *Carbon and Climate Law Review, 13*(1), 31–49. doi:10.21552/cclr/2019/1/6

Rajamani, L. (2010). The increasing currency and relevance of rights-based perspectives in the international negotiations on climate change. *Journal of Environmental Law, 22*(3), 391–429. doi:10.1093/jel/eqq020

Rajamani, L. (2018). Human rights in the climate change regime. In *The human right to a healthy environment* (pp. 236–251). Cambridge: Cambridge University Press.

Rogelj, J., den Elzen, M., Höhne, N., Fransen, T., Fekete, H., Winkler, H., … Meinshausen, M. (2016). Paris Agreement climate proposals need a boost to keep warming well below 2 degrees C. *Nature, 534*(7609), 631–639. Retrieved from http://www.ncbi.nlm.nih.gov/pubmed/27357792

Savaresi, A. (2018). Climate change and human rights. Fragmentation, interplay and institutional linkages. In S. Duyck, S. Jodoin, & A. Johl (Eds.), *Routledge Handbook of Human Rights and climate Governance* (pp. 32–42). London: Routledge.

Savaresi, A., & Hartmann, J. (2015). *Human rights in the 2015 agreement.* Retrieved from https://legalresponse.org/wp-content/uploads/2015/05/LRI_human-rights_2015-Agreement.pdf

Sheeran, S. (2014). The Relationship of international human rights law and general international law: Hermeneutic constraint, or pushing the boundaries? In *Routledge handbook of international human rights law* (pp. 95–124). New York: Routledge.

SIDS. (2007). *Malé declaration on the human dimension of global climate change.* Retrieved from www.ciel.org/Publications/Male_Declaration_Nov07.pdf

Toussaint, P. (2019). Voices unheard – affected communities and the climate negotiations on loss and damage. *Third World Thematics: A TWQ Journal,* 1–20. Retrieved from https://www.tandfonline.com/doi/full/10.1080/23802014.2018.1597640

UNEP. (2015). *Climate change and human rights.* Nairobi. Retrieved from http://columbiaclimatelaw.com/files/2016/06/Burger-and-Wentz-2015-12-Climate-Change-and-Human-Rights.pdf

UNFCCC. (1992). *United nations framework convention on climate change.*

UNFCCC. (1995). *Report of the conference of the parties on its first session, held at Berlin from 28 March to 7 April* 1995. FCCC/CP/1995/7. Retrieved from http://unfccc.int/cop5/resource/docs/cop1/07.pdf

UNFCCC. (2008). *Report of the conference of the Parties on its thirteenth session, held in Bali from 3 to 15 December* 2007. FCCC/CP/2007/6. Retrieved from https://unfccc.int/resource/docs/2007/cop13/eng/06.pdf

UNFCCC. (2014). *Initial two-year workplan of the executive committee. In section III.C. of the report of the executive committee of the Warsaw international mechanism for loss and damage associated with climate change impacts.* FCCC/SB/2014/4. Retrieved from https://unfccc.int/resource/docs/2014/sb/eng/04.pdf

UNFCCC. (2017a). *Submission from the advisory group on climate change and human mobility on possible activities under strategic workstreams of the five-year rolling workplan.* Retrieved from https://unfccc.int/files/adaptation/groups_committees/loss_and_damage_executive_committee/application/pdf/advisory_group_submission.pdf

UNFCCC. (2017b). *Submission on the strategic workstream on loss and damage finance. To the executive committee of the Warsaw international mechanism for loss and damage on behalf of climate justice justice programme, Heinrich Böll Stiftung, stamp out poverty.* Retrieved from https://unfccc.int/files/adaptation/groups_committees/loss_and_damage_executive_committee/application/pdf/climate_justice_programme_heinrich_boell_stamp_out_poverty.pdf

UNFCCC. (2018). NDC registry. Retrieved from https://www4.unfccc.int/sites/ndcstaging/Pages/Home.aspx

UNGA. (1970). 2625 *(XXV). Declaration on principles of international law concerning friendly relations and co-operation among States in accordance with the charter of the United Nations.* A/RES/25/2625. Retrieved from http://www.un-documents.net/a25r2625.htm

UNGA. (2007). *Right of everyone to the enjoyment of the highest attainable standard of physical and mental health.* A/62/214. Retrieved from http://undocs.org/A/62/214

UNGA. (2015). *Resolution 70/1. Transforming our world: the 2030 agenda for sustainable development.* A/RES/70/1. Retrieved from https://undocs.org/A/RES/70/1

UNHRC. (2008). *Resolution 7/23. Human rights and climate change* A/HRC/RES/7/23. Retrieved from http://ap.ohchr.org/documents/e/hrc/resolutions/a_hrc_res_7_23.pdf

UNHRC. (2009a). *Report of the Office of the United Nations High Commissioner for Human Rights on the relationship between climate change and human rights.* A/HRC/10/61. Retrieved from http://undocs.org/A/HRC/10/61

UNHRC. (2009b). *Resolution 10/4. Human rights and climate change.* A/HRC/RES/10/4. Retrieved from https://ap.ohchr.org/documents/E/HRC/resolutions/A_HRC_RES_10_4.pdf

UNHRC. (2016). *Report of the Special Rapporteur on the issue of human rights obligations relating to the enjoyment of a safe, clean, healthy and sustainable environment.* A/HRC/31/52. Retrieved from https://www.ohchr.org/EN/HRBodies/HRC/RegularSessions/Session31/Documents/A%20HRC%2031%2052_E.docx

UNHRC. (2017a). *Analytical study on the relationship between climate change and the full and effective enjoyment of the rights of the child.* A/HRC/35/13. Retrieved from https://undocs.org/en/a/hrc/35/13

UNHRC. (2017b). *Resolution 35/20. Human rights and climate change* A/HRC/RES/35/20. Retrieved from https://www.right-docs.org/doc/a-hrc-res-35-20/

UNHRC. (2018a). *Human Rights Council holds panel discussion on climate change and the rights of the child*. Retrieved from https://www.ohchr.org/EN/HRBodies/HRC/Pages/NewsDetail.aspx?NewsID=22757&LangID=E

UNHRC. (2018b). *Report of the Special Rapporteur on the issue of human rights obligations relating to the enjoyment of a safe, clean, healthy and sustainable environment* A/HRC/37/59. Retrieved from https://undocs.org/A/HRC/37/59

UNHRC. (2018c). *Report of the Special Rapporteur on the right to food*. A/HRC/37/61. Retrieved from http://undocs.org/A/HRC/37/61

UNHRC. (2018d). *The Slow onset effects of climate change and human rights protection for cross-border migrants*. A/HRC/37/CRP.4. Retrieved from https://www.ohchr.org/Documents/Issues/ClimateChange/SlowOnset/A_HRC_37_CRP_4.pdf

Verheyen, R., & Roderick, P. (2008). Beyond Adaptation: The legal duty to pay compensation for climate change damage. *WWF-UK Climate Change Programme discussion paper*.

Walker, S. (2005). Human rights impact assessments of trade-related policies. *Gehring M., Cordonier-Segger M. Claire. Sustainable Development in World Trade Law. Kluwer: Law*, 217–256.

Warner, K., Van der Geest, K., Huq, S., Harmeling, S., Kusters, K., de Sherbinin, A., & Kreft, S. (2012). *Evidence from the frontlines of climate change: Loss and damage to communities despite coping and adaptation*.

Loss and damage: an opportunity for transformation?

Erin Roberts and Mark Pelling

ABSTRACT

As observed and predicted losses and damages from climate change impacts grow increasingly severe, calls for transformation as a response to long-term climate change have become more frequent. Transformational approaches have also been integrated into the global climate change regime under the UN Framework Convention on Climate Change (UNFCCC) as part of the workplan of the Executive Committee guiding the implementation of the Warsaw international mechanism, the oversight body on loss and damage. However, there has as yet been no attempt to define what is meant by transformation in the context of loss and damage. This paper attempts to clarify the burgeoning academic and policy literature by positing three types of transformation as a response to loss and damage: transformation as adaptation (an intensification of dominant socio-ecological relationships), transformation as extension (when the limits of established adaptive capacity are reached) and transformation as liberation (adopting development pathways that address the root causes of vulnerability). Transformation as liberation is proposed as a deeper change to social-technological systems to avoid and minimize loss and damage in ways that enhance social justice and sustainability. To provide the kind of information decision makers need to plan and implement transformation as liberation, more research is needed on how to plan in a way that ensures the most equitable outcomes.

Key policy insights

- Loss and Damage is an opportunity to scrutinize and address the root causes of vulnerability.
- Framing climate change as a development crisis will allow opportunities for transformation as liberation to emerge.
- Transformation as liberation to address the root causes of vulnerability requires meaningful engagement with processes at all levels.
- A new model of global governance is needed in which global equity is a moral imperative.
- The transition to transformation as liberation must be just, which requires leadership, inclusive and participatory decision making and building alliances.
- The global Loss and Damage agenda could open up space for a broader discussion on how transformation as liberation can be facilitated to address inequalities both between and within countries.

Introduction

Over the past few years, extreme weather and slower onset climatic events have resulted in unprecedented levels of loss and damage in countries around the globe. With the impacts of climate change on the rise, calls for transformational approaches to address the climate challenge have become increasingly frequent (Barnett et al., 2015; Eriksen, Inderberg, O'Brien, & Sygna, 2015; Klein et al., 2014; Mapfumo et al., 2015; Moore et al., 2014; O'Brien, 2012; 2017; Pelling, 2011; Warner & van der Geest, 2013; Westley et al., 2011,

2013). Transformation has become associated with loss and damage and both have become synonymous with the limits to adaptation. Dow, Berkhout, Preston, Klein, et al. (2013) assert that when the limits to adaptation are reached, a society has two choices: incur loss and damage, or transform. Yet, although transformation is a term increasingly employed within the global discourse on climate change, it remains ambiguous with multiple interpretations (Feola, 2015; Nalau & Handmer, 2015).

Though Parties to the Paris Agreement agreed to the importance of averting, minimizing and addressing losses and damages from the impacts of climate change, there is no universally accepted definition of loss and damage. Climate scholars refer to both dimensions broadly as the negative impacts of climate change (Kreft, Warner, Harmeling, & Roberts, 2013; UNFCCC, 2012), and more narrowly as the impacts of climate change that are not avoided by mitigation and adaptation (Roberts & Pelling, 2018). Surminski and Lopez (2015) propose that framing loss and damage has both a technical dimension which focuses on the practical aspects of avoiding, minimizing and addressing loss and damage and a political dimension with its focus on historical liability and compensation. To differentiate between the impacts and the tools used to address them, we use 'loss and damage' (lower case) to refer to the impacts of climate change not avoided by mitigation and adaptation efforts and 'Loss and Damage' (upper case) to refer to the broader policy frameworks at all levels within which these impacts are being addressed.

The association of transformation with Loss and Damage arises from academic work and has also been recognized by the UNFCCC under the Warsaw International Mechanism on Loss and Damage (WIM), the overarching body on Loss and Damage within the global climate regime (UNFCCC, 2017). In their research amongst key stakeholders, Boyd, James, Jones, Young, and Otto (2017) identified the consideration of transformation in development or risk management pathways once residual impacts in vulnerable countries become unacceptable as one aspect of the global Loss and Damage agenda. This perspective is contrasted with others which include the need for scaling up both mitigation and adaptation efforts, the importance of comprehensive risk management and evidence of the limits to adaptation. In practice, these perspectives overlap and could reinforce each other in determining policy orientation, ambition and application to address the root causes of vulnerability (Roberts & Pelling, 2018). Despite limited engagement with scholarly debates, the evolution of Loss and Damage has also come to include statements on transformation. The global Loss and Damage agenda under the UNFCCC is being used as grounds to call for transformative responses to address climate change and limit its impacts (Roberts & Pelling, 2018; Roberts, van der Geest, Warner, & Andrei, 2014; Warner & van der Geest, 2013; Warner et al., 2012). Transformational approaches to address loss and damage have been included in the workplans of the Executive Committee, which guides the work of the WIM – the oversight body on Loss and Damage under the UNFCCC. The term 'transformational approaches' has yet to be defined within these discussions, however, as it remains contentious.

Transformation is not a novel concept and has been discussed in various fields of study for decades (see for example: Freire, 1970). As the climate crisis broadens and deepens transformation is experiencing a resurgence but in many ways these debates are age-old. This paper is motivated by the emergence of the Loss and Damage agenda as a platform for transformation, and the need for a systematic assessment of the ways in which transformation might be deployed to best meet the aims of Loss and Damage policy to avert, minimize and address the impacts of climate change that are not avoided. These include reducing the impacts of climate change while meeting the goals of equity and sustainability as agreed in sister frameworks, importantly the UN Sustainable Development Goals (SDGs) and the Sendai Framework for Disaster Risk Reduction. Policies to avert, minimize and address loss and damage encompass both transformative – fundamental – changes in technical dimensions of risk management and the broader economic, social and political structures that underpin climate vulnerability and impacts. This paper begins by demonstrating the range of existing interpretations of transformation that can, or have been, applied to shape Loss and Damage policy and practice. Three approaches are identified, with overlapping features: transformation as intensification, transformation as extension and transformation as liberation. The paper analyses these approaches, and concludes that transformation as liberation offers the widest range of policy opportunities for Loss and Damage to meet the goals of equitable and sustainable development. The paper then offers a set of potential enabling factors and recommendations to help transition transformation policy to practice and also highlights the role of global processes in facilitating transformation.

Unpacking transformation as a response to increasing loss and damage

Transformation as a response to climate change impacts has origins in many disciplines (Bahadur & Tanner, 2012; Pelling, 2011). Within the climate change adaptation literature, notions of transformation first arose in work on social-ecological systems (Gunderson & Holling, 2002; Gunderson, Holling, & Light, 1995; Holling, 1986; Walker, Holling, Carpenter, & Kinzig, 2004). In their seminal paper, Walker et al. (2004, p. 2) defined transformability as '[t]he capacity to create a fundamentally new system when ecological, economic, or social (including political) conditions make the existing system untenable'. The social-ecological systems literature differentiates between transformation in which a regime shift occurs and incremental adaptation in which the regime remains intact (Folke, 2006; Olsson, Galaz, & Boonstra, 2014 in Fenton, Tallontire, & Paavola, 2016). The relationships between transformation and resilience continue to be debated depending on whether transformation is seen as a process or an outcome. For Walker et al. (2004), a resilient system is able to absorb and respond to shocks while transformation occurs when the limits of adaptability are reached and a system can no longer maintain its resilience. Pearson and Pearson (2012) propose that resilience tends to be inward looking, aimed at maintaining a system, whereas transformation is more outward looking, and more open to new ways of thinking and doing. However, Pelling, O'Brien, and Matyas (2015) argue that, within adaptation policy, transformation can lead to both resilient and sustainable development outcomes. They also argue that incremental adjustments can similarly lead to resilient outcomes, though Matyas and Pelling (2014) maintain that incremental changes can also be short-term fixes that delay or avoid entirely more fundamental changes, and can lock systems or societies into unsustainable pathways.

Following, in particular, the Special Report of the Intergovernmental Panel on Climate Change (IPCC) on Managing the Risks of Extreme Events and Disasters to Advance Climate Change Adaptation (SREX) (IPCC, 2012), views on transformation inspired by political economy and political ecology were introduced to climate change adaptation. Developed within the field of disaster risk reduction, these complemented social vulnerability approaches to climate change adaptation (Adger, 1999). Within the tradition of political economy, transformation indicates a change to social, political or economic structures with the depth and breadth of change differing across the spectrum of concepts of transformation (Feola, 2015; Godfrey-Wood & Naess, 2016). Building on these literatures, Fenton et al. (2016) proposed three categories of transformation in the context of climate change: the ecological resilience approach based on the understanding of transformation in the social-ecological systems literature; the political ecology approach in which transformation challenges the status quo and addresses the root causes of vulnerability; and, finally, a risk hazard approach, which views adaptation at larger scales and intensities. An underlying theme in all three categories is that, for transformation to occur, the status quo must be untenable and the proposed transformation must be novel practice (Fenton et al., 2016). In this section, we acknowledge from the outset the complexities of defining transformation, and that responses to climate change are inherently political and shaped by those in power (see: Adger, 2003; Eriksen et al., 2015; Pelling, 2011). Our aim is to draw a link between the global climate change agenda and national and sub-national policies and processes to suggest how Loss and Damage could provide the impetus for broader and deeper change.

Transformation as intensification of dominant social-ecological relationships

Transformation as intensification of dominant social-ecological relationships reflects two types of adaptation. One is based on the theory that incremental adaptation strategies can build on one another and eventually lead to transformational change. The second is that technological changes can by themselves be characterized as transformation. These types of transformation are defined as an intensification of dominant social-ecological relationships as they reinforce rather than challenge the status quo. With warming of over 2°C increasingly likely, adaptation will need to cycle between incremental and transformative actions (Park et al., 2012; Wise et al., 2014). In fact, some scholars have proposed that incremental adaptation can also lead to transformative outcomes (Mapfumo et al., 2015; Park et al., 2012; Rippke et al., 2016). However, the extent to which these outcomes are indeed transformational is debated.

Much work on transformational adaptation has focused on the agricultural sector. Rippke et al. (2016) describe as transformational a process which begins with incremental adaptation including improvements to crops and practices, proceeds to a preparatory phase to develop and implement enabling institutions and policies, and ends with what they refer to as transformational adaptation, in which farmers grow new crops, explore other livelihood strategies or relocate. However, these adaptation strategies do not challenge the status quo and describe incremental changes in response to changing conditions. This is more akin to business as usual adaptation than transformation. Similarly, Rickards and Howden (2012) describe transformational adaptation in agriculture as expressions of ongoing co-evolution. In their depiction, adaptation is deployed as a technology to respond to climatic changes, not to address the underlying political, social, cultural, and economic conditions that give rise to vulnerability. In this understanding, transformation is applied to describe adaptation leading to an intensification of the status quo. The fittest will survive and may prosper through adaptation, while those less able are increasingly left behind or negatively impacted by the adaptation choices of others (Atteridge & Remling, 2017).

The second type of transformation as intensification is described by Kates, Travis, and Wilbanks (2012), who define transformation as adaptation that takes place at a larger scale or intensity than previous adaptation actions, adaptation that takes place at a different place or location (such as resettlement) and new types of adaptation, such as the adoption of new climate-resistant crop varieties. There is no explicit acknowledgement of the opportunity – or responsibility – opened by a desire to avoid future loss and damage by realigning underlying development trajectories towards unmet sustainable development goals. Intentional change under transformation as intensification is limited to technology, though secondary or unintended consequences may also transform ecological or social realms. In Australia, Marshall, Park, Adger, Brown, and Howden (2012) have used transformation to describe economic intensification processes and find them preferred by actors who are well-networked and better resourced.

Outcomes intensify dominant economic and social-ecological relations and tend to benefit already powerful actors while potentially inflicting secondary impacts – or losses and damages – on those whose livelihoods are dependent upon them. Also working in Australia, Park et al. (2012) demonstrate this effect in the wine sector, with well-financed businesses adapting in ways described as transformative by changing location, technology or organizational form, with knock-on implications for host communities and supply chain partners left behind. Thus, transformation understood as intensification reproduces patterns of inequality, and access to opportunities to transform are unequal. Thornton and Camberti (2013, p. 4) take a slightly different view, calling transformational adaptation, 'the radical end of more conventional adaptation processes and capacities'. One of the examples they provide is the development of renewable energy initiatives in Alaskan indigenous communities as a means of reducing dependence on fossil fuels, creating employment and maintaining rural livelihoods. In this sense transformation as intensification is more empowering and has the potential to build resilience and avoid and minimize losses and damages in the long-term by enhancing sustainable development. However, the intention is not to fundamentally change social systems. These examples of incremental adaptation by themselves are unlikely to lead to transformation unless they inspire fundamental changes that address unequal balances of power that allow some to employ adaptation strategies not available to others.

Clearly, adaptation has a role to play in avoiding and minimizing losses and damages, but incremental adaptation will not be sufficient to avoid loss and damage in all cases (Klein et al., 2014). In the context of Loss and Damage, transformation as intensification has a role to play but raises questions of equity as not all actors will have the capacity to implement transformation as intensification. Transformation understood as intensification does not fundamentally alter existing relationships nor does it address the reasons why households, societies and systems are vulnerable in the first place. This constrains the scope for contributions to enhance sustainable and equitable development. We would therefore not characterize transformation as intensification as true transformation.

Transformation as extension of the limits to adaptation

The second meaning ascribed to transformation of relevance to the Loss and Damage agenda is implemented when the limits to adaptation are approached or have been reached. We refer to this transformation as an extension of the limits to adaptation. Both Dow, Berkhout, Preston, Klein, et al. (2013) and Preston, Dow, and Berkhout (2013) assert that, when the limits to adaptation are reached, either transformation occurs or losses and damages are incurred. That said, transformation can also result from the manifestation of loss and damage. Deliberate transformations that are planned when the limits to adaptation are reached involve the re-orientation of the objectives and priorities of a system (Preston et al., 2013). This interpretation shares with transformation as intensification a conservative orientation. Neither seeks to consider nor purposefully engage with underlying development structures, though both will have local impacts on both development opportunities and outcomes.

Transformation as extension is largely characterized in the literature as a spontaneous process which often exacerbates inequality and deepens poverty and vulnerability. Nelson, Adger, and Brown (2007) describe this variant of transformation as the point at which the ecological, social or economic conditions of a system become untenable or undesirable. In this sense, either collapse or transformation are forced upon a system or society. The presentation of transformation as extension echoes work on coping with disaster loss and especially food insecurity (Wisner, Cannon, Davis, & Blaikie, 2004). Here coping is presented as a set of cascading systems transformations (Pelling, 2011), with each transformation reducing future opportunities as households or individuals are forced to make trade-offs that will ultimately make them more vulnerable to future impacts. Each transformation allows the household to extend its survival in the near term, but at a cost to long-term resilience (see: Opondo, 2013; Warner & van der Geest, 2013).

Research in Australia described the way in which a peanut company responded when it recognized the limits of incremental adaptation had been reached by buying land in another part of the country with access to irrigation (Jakku et al., 2016). Ultimately, however, economic, environmental, social and institutional barriers were encountered and eventually the company shifted from raw peanut production to value-added processing. In fact, in many ways, this example is more akin to transformation as intensification than transformation as extension and the definition of transformation as proposed by Kates et al. (2012). We characterize it as transformation as extension as it represents a deliberate effort to extend the limits of adaptation although it reinforces rather than challenges the status quo. Similarly, Fenton et al. (2016) found that wealthier households in rural Bangladesh implemented transformation as extension by purchasing land from poorer households to expand aquaculture operations when traditional agriculture became untenable. In both examples resource levels determined the ability of actors to adapt and avoid or reduce loss and damage.

Transformation as extension may extend the limits to adaptation of some, but those who benefit from these efforts are those with the most resources. The limits to adaptation are heavily influenced by the values underlying adaptation preferences – be they to avoid intolerable risk or to improve human well-being as part of a broader development process – and can be difficult to assess (Adger et al., 2009). The use of transformation to extend the limits of adaptation recognizes the complexity of the limits to adaptation. First and foremost, the limits to adaptation are fundamentally about the value and what level of risk can be tolerated. The discourse on the limits to adaptation opens up a discussion on how the limits to adaptation can be extended which is relevant for the loss and damage agenda. From the literature, it is clear that societies will need to make some difficult choices when the limits of adaptation are reached in order to avoid and minimize loss and damage. And in fact, avoiding loss and damage in one context may give rise to losses and damage in another. In order to understand who will benefit from transformation as extension, a greater understanding of the prevailing power dynamics in society is required. Current examples of transformation as extension in the literature are not, in our view, transformation.

Transformation as liberation to address the root causes of loss and damage

Transformation as liberation describes efforts to enhance equity and justice as a means of reducing vulnerability and risk and as contributions to wider and ongoing development struggles. The extent of losses and damages arising from climate-related hazards is influenced by the vulnerability – be it social, political or economic – of the people and societies exposed to them (Ribot, 2010). Recognizing this, many scholars therefore see transformation in the context of climate change as a deeper, more challenging change, increasingly considering issues of social justice and imbalances of power (Moore et al., 2014). We refer to this type of transformation as liberation to address the root causes of climate change because it provides for a scrutinization of the forces that drive the marginality and inequality that contribute to vulnerability. Béné, Newsham, Davies, Ulrichs, and Wood (2014, p. 22) argue that transformation involves challenging the status quo by altering 'entrenched systems maintained by powerful interests'. This liberatory understanding of transformation retains dependency for change on a climate event. For Béné et al. (2014) a crisis is needed to provoke transformation. However, transformation can be both unplanned and inadvertent or planned and deliberate and can have both positive and negative outcomes (Feola, 2015; Wilson, Pearson, Kashima, Lusher, & Pearson, 2013).

Manuel-Navarrete and Pelling (2015) define deliberate transformation as radical changes that can simultaneously reduce both inequality and climate risks. Matyas and Pelling (2014, p. 58) go one step further, defining transformation as 'fundamental restructuring, pushing the system towards a different status quo'. These changes can either be purposefully planned and carried out by human actors or triggered by an external event (Manuel-Navarrete & Pelling, 2015). We argue that transformation as liberation is transformation that is deliberate and planned, and which engages with those most vulnerable. However, it must be acknowledged that transformation as liberation is difficult to achieve (Fenton et al., 2016). Can liberation in adaptation also be anticipatory and catalyzed not by an event but by risks and the evidence they provide of unsustainable development pathways? If so climate change with its unique futures orientation offers a significant opportunity for social liberation that can also reduce loss and damage by bringing together justice and resilience.

Transformation as liberation is defined by its capacity to understand and address what makes households and societies vulnerable. Pelling et al. (2015) maintain that transformation occurs when existing systems and decision makers aim to address the root causes of social vulnerability. In this way, transformation as liberation can correct the shortcomings of historic development pathways that have created or contributed to the social inequality, poverty and environmental problems eventually leading to vulnerability (Eriksen et al., 2011). Transformation as liberation addresses development failures including by extending rights to those previously marginalized. Given that it challenges the status quo, transformation is more likely to be driven by a bottom up than a top down process (Manuel-Navarrete & Pelling, 2015). Transformation as liberation is not only a deeper change but also a longer-term strategy that can avoid and minimize loss and damage and provide a platform to better address losses and damages when they occur by ensuring a wider range of actors have the tools they need. It is in increasing informed and inclusive decision-making with the power to act that adaptation can contribute to more just processes and outcomes and become liberatory (Ziervogel et al., 2017).

Transformation as liberation can also fundamentally alter the way in which people see themselves and their relationships to both society and others within it. For O'Neil and Handmer (2012, p. 2), transformation is a call for re-examining the way in which people relate to one another, their environment and broader social processes. In this understanding of transformation as liberation, both the risk and manifestation of loss and damage open up scope for reflecting on, and changing, perceptions, meaning, norms and values, altering social networks and power structures, and introducing new institutions and regulatory frameworks (O'Brien et al., 2012). Transformation as liberation requires taking a broader set of issues and ideas into consideration alongside climate change policies and plans, and questioning current development trajectories. This requires an understanding of vulnerabilities and a vision for a preferred future. This vision of transformation is emerging in practical application. The opening of space for development gains through adaptation planning can be seen in a recently developed African Development Bank programme that provides access to both land and credit for women (AfDB, 2015) to confront women's lack of land tenure rights, which is one of the predominant causes of gendered inequality in vulnerability to climate change and wider development opportunity (ActionAid, 2011).

Transformation in global processes

Climate change is fundamentally a development crisis (Francis, 2015; Parry, 2009). If approaching transformation as liberation offers the best hope of transitioning towards a more sustainable and equitable future, then what scope is there for the evolving UNFCCC mechanisms to support this process? Within the UNFCCC architecture the WIM was established at the 19th Conference of the Parties (COP) in 2013 to address loss and damage in developing countries particularly vulnerable to the impacts of climate change (UNFCCC, 2014). The workplan of the Executive Committee that guides the implementation of the WIM explicitly identifies transformational approaches in the context of comprehensive risk management (UNFCCC, 2014). However, the term 'transformational approaches' have yet to be defined and remain politically contested.

While transformation has different meanings for different actors, the lack of an agreed definition within the UNFCCC could allow the term to be co-opted by powerful actors who will impose their own definition of trans-formation (Feola, 2015). Moreover, under the UNFCCC negotiations, transformation is highly political, often favoured by developed country actors, perhaps as a substitute for providing support to compensate developing countries for loss and damage. A similar process has unfolded across IPCC reports. While transformation had a singular interpretation – fundamental change - when first introduced in the 2012 SREX (IPCC, 2012), the three competing interpretations presented in this paper were expressed only two years later in different chapters of the IPCC Fifth Assessment Report (IPCC, 2014; Pelling, 2014).

There are additional challenges for the global agenda. Transformation as extension could potentially under-mine the Loss and Damage agenda. For example, if relocation – be it of a community or the citizenry of an entire country - is framed as transformation, as in the IPCC's Fifth Assessment Report, the argument could be that there is no need to develop approaches to address loss and damage, as transformational adaptation is sufficient to respond to the impacts of climate change. However, if framed as a means of avoiding loss and damage, trans-formation could be used to extend the limits of adaptation in a meaningful way if it addressed the underlying power structures that contribute to the conditions that give rise to vulnerability. Whether or not an action is framed as a transformation or an approach to address loss and damage – as with relocation – would also depend on the objectives and values of the society taking that action. If the objectives of a society continue to be met, then it has not yet reached the limits of adaptation (see Dow, Berkhout, & Preston, 2013; Preston et al., 2013). For example, if planned in such a way that it moves societies towards more resilient, equitable and resilient futures, relocation could in fact be transformative. If, however, the objectives of a society are no longer met and they are forced to take action to avoid intolerable risk, then it could be said they are already incurring loss and damage while taking action to avoid greater losses and damages. However, this also requires answering the question of who gets to decide when the limits to adaptation are reached. If it is the powerful and elite, and the status quo is perpetuated, this could exacerbate the vulnerability of the poorest and most marginalized.

Under the UNFCCC process, there are already provisions for increasing focus on people and countries that are particularly vulnerable to the impacts of climate change. The adaptation article of the Paris Agreement acknowl-edges the importance of adaptation action being country-driven, gender-responsive, participatory and fully transparent and taking into account vulnerable groups, communities and ecosystems (UNFCCC, 2016, Article 7, paragraph 5). If this was truly achieved, from the local to the global levels, it would amount to a transformation of decision-making processes and an advance for procedural justice – transformation as liberation to avert, mini-mize and address loss and damage. Transformation as liberation also provides opportunities to scrutinize some of the processes that render certain populations vulnerable. However, there are vested interests maintaining powerful narratives that sustain the status quo (Joshi, Platteeuw, Singh, & Teoh, 2018; O'Brien et al., 2012; Pelling, 2011; Pelling et al., 2015). Using the example of Bangladesh, Paprocki (2015) argues that the narrative of climate change can depoliticize development and obscure some of the political and social injustices that have exacerbated in equality. This highlights the challenge of transformation as liberation as a top down process given that it is often in the interests of the most powerful in society to maintain the status quo.

Transformation as liberation requires focus on the processes that render countries, societies and households vulnerable at all levels. Both the UNFCCC and its Paris Agreement have re-enforced the sovereign right of each national state to determine its own adaptation strategies. It may be difficult for alternative development models to be envisioned or deployed even as experiments. However, what if both the vulnerability of countries within the global regime under the UNFCCC and that of the marginalized populations within them could be simul-taneously recognized and addressed? What if developing countries could advocate for greater resources to address the impacts of climate change for which they are not responsible while simultaneously addressing the root causes of vulnerability within their own borders? Increasing focus on reducing vulnerability could provide an entry point and increased legitimacy for sub-national actors and NGOs working on the ground to support transformation as liberation. This is already happening, for example through the work of international NGOs like ActionAid and CARE that have integrated tools for empowerment into climate change programming (see: ActionAid, 2014; CARE, 2015). The challenges remain the extent to which alternatives and dominant approaches can coexist with one another and the consequences of arising social and political tension. Ensuring a meaningful discussion on transformation as liberation within the global climate regime would require a broader and deeper conversation about why developing countries are vulnerable in the first place and why they continue to lack an equitable voice within global processes.

There are also entry points for transformation as liberation in the global sustainable development agenda. In June 2015, Pope Francis issued a 180-page encyclical which situated modern capitalism at the heart of the 'climate crisis'. The encyclical called for an altering of production and consumption patterns in developed countries to reduce future climate change impacts and address the root causes of poverty in developing countries, drawing links between the two (Francis, 2015). The UN SDGs share a vision of coupling social justice and resilience, and provide a more technical entry point to compliment Pope Francis's normative leadership. While SDG 13 is explicitly focused on addressing climate change, all the SDGs could enable the opening of broader and alternative development visions and directions, particularly SDG 5 (achieving gender equality and empowerment of women and girls), SDG 10 (reducing inequality within and among countries) and SDG 12 (ensuring sustainable production and consumption patterns). Again, it is up to each country to implement strategies to address vulnerability and reduce poverty in their own national context, but given their focus on empowerment and equality, the SDGs present an opening for transformation as liberation to avoid, minimize and address loss and damage through more just development. Again, however, transformation as liberation requires acknowledging the root causes of vulnerability at all levels.

Planning for and implementing transformation as liberation within the loss and damage agenda

As the intended outcome of a planned process or the unplanned outcome of a spontaneous process or event (see: Moore et al., 2014; Nelson et al., 2007; O'Brien, 2012; Pelling et al., 2015), transformation as liberation can evolve slowly as a result of changing social values and institutions, or quickly as a result of an external event such as a disaster (O'Brien, 2012; Pelling, 2011). Exogenous shocks can provide opportunities for re-evaluating the status quo (Folke et al., 2010) and open up space for transformation as liberation (Pelling & Dill, 2010). Part of the response to the 2010 earthquake in Christchurch, New Zealand, for example, has been to open new educational and small-business support opportunities for Maori youth, and this has spread to nationwide application (Moore et al., 2014). Hurricane Sandy increased awareness of climate change among individuals and in public and political discourse, reinforcing mitigation and transformative approaches to adaptation in New York City (Rosenzweig & Solecki, 2014). With support, similar approaches could be adopted in the Caribbean and southern Africa, regions that will be rebuilding for years to come in the wake of recent cyclones.

A significant challenge to transformation as liberation is that it involves fundamental changes that themselves bring social impacts, often initially to the poorest and most vulnerable. Transformation requires understanding and challenging the assumptions underlying prevailing development discourses and practices (Manuel-Navarrete & Pelling, 2015). Key to transformation is the question of who decides when transformation is needed and how it unfolds (O'Brien, 2012). Transformation as liberation acknowledges that policies and plans to avert, minimize and address loss and damage can be captured by already dominant interests to accelerate the status quo reducing options for just, sustainable and resilient futures. As O'Brien (2017) notes, transformation is not a neutral concept, but one which is reflective of the beliefs, values, worldviews and interests of both actors and institutions. Because transformation as liberation involves intentional and fundamental social change within systems maintained and protected by powerful actors, attempts to foster transformation will likely be hidden from the view of dominant actors (including science) (O'Brien, 2012; Pelling, 2011; Pelling & Manuel- Navarrete 2011).

Transformation as liberation is rare as it requires changes to systems maintained by powerful actors (Béne et al., 2014) and tends to be followed (and also often proceeded) by a period of instability which development and humanitarian agencies are uncomfortable with provoking (Pelling et al., 2015). Stable societies are organized to resist transformative tendencies (whatever their direction or cause) and reaching beyond stability opens many uncertainties for all segments of society but especially for the poorest who have the least resource to cope with change (Pelling, 2011). Transformation thus carries a heavy ethical load – those who may benefit most in the long-run might also lose most in the short-run. This raises strong questions for the development of policies and plans to avert, minimize and address loss and damage and facilitate sustainable development. It is therefore important to understand the 'disequilibria' that can result from transformation (O'Brien et al., 2012, p. 466). As Fook (2015) argues, transformation poses more questions than it can answer – though this could be characterized as an opportunity rather than a drawback. Indeed, transformation forces us to look at ourselves and our societies, including the critical links between countries and the globalizing forces that give rise to vulnerability. Ensuring adequate lead-time for transformation can smooth the transition between incremental change and transformation as liberation (Stafford, Horrocks, Harvey, & Hamilton, 2011) and prevent possible problems before they arise (Rickards & Howden, 2012).

What are the precursors of transformation as liberation?

When transformation as liberation does occur, it is often the result of a pressure from several fronts which cut across scales (Smith, Stirling, & Berkhout, 2005 in Moore et al., 2014). Leadership is also an important component of planning for transformation and can bring these elements together. Moore et al. (2014) argue that, regardless of who is leading, establishing collective vision for alternative development pathways and the gathering of momentum for change are important precursors to transformation. Empirical research in Australia found that an ability to develop a vision of alternative futures is a precursor to transformation (Wilson et al., 2013). Though more ambitious climate change policies at the national level can stimulate greater action at the local level, local leadership is also important (Burch, 2010). In their analysis of eight case studies in Africa, Mapfumo et al. (2015) found that 'change agents' are important precursors of transformational change. However, even if the political leadership exists, transformation as liberation is unlikely to be realized unless individuals have the necessary knowledge, determination and capacity to affect change (Räthzel & Uzzell, 2011). Self- and critical- reflection (Schlitz, Vieten, & Miller, 2010) and how individuals see themselves, their relationships to others and the environment are also key to planning for deliberate and less disruptive transformation as liberation (Pelling, 2011).

Inclusive and participatory decision making that responds to a range of values and objectives is also a crucial first step towards transformation that could be liberatory. Vulnerable groups are often excluded from decision-making processes yet forced to live with their outcomes (Adger, 2003). However, participation alone is not enough if it does not acknowledge local power dynamics (Dodman & Mitlin, 2013). There are many well-documented examples of large social organizations, especially in South and Southeast Asia, that have leveraged social change within alternative development paradigms from community-based origins (Mitlin, 2012). One example of this is in the work of the Bangladesh Rural Advancement Committee to enhance social, legal and political awareness of rural women in Bangladesh, with the aim of helping women transition from being dependent on their husbands to earning money of their own (Hashemi & Umaira, 2011). A focus on transparency, accountability and empowerment provides a space for disempowered groups to have a voice in decision making while enhancing their ability to hold those in power responsible for their decisions (Ensor, Park, Hoddy, & Ratner, 2015). In the wake of the 2010 floods in Pakistan 1000 women leaders were trained to better understand their rights and subsequently created self-help groups in their own communities, one of the aims of which is to negotiate with local governments to develop more effective risk reduction planning that better integrates the needs of women and girls (Action Aid, 2014). It is difficult to measure the impacts of these policies and programmes as losses and damages not incurred cannot be measured. In addition, as Parsons and Nalau (2016) note, the true judges of whether or not transformation has occurred is those whose lives have been transformed (or not).

Can local transformations lead to systems-wide change? Work on social movements suggests that social tipping points do not require action support from the majority of society. While learning and a willingness to experiment are also essential for transformation (Mapfumo et al., 2015; Moore et al., 2014; Pelling, 2011) local experiments are unlikely to generate transformation beyond individual and local experience, without behavioural changes at larger scales (Tschakert & Dietrich, 2010). These large scale behavioural changes could be prompted with a re-framing of climate change from an environmental problem to a social and developmental crisis. Sustaining transformation will also require building alliances and networks – both social and institutional –

across both scales and sectors (Mapfumo et al., 2015). A strong sense of community identity can also help facilitate transformation (Wilson et al., 2013).

Whether or not transformation of any kind occurs has a lot to do with how climate change is framed (Rickards, 2013). If it is isolated as a purely environmental problem rather than a result of unsustainable development pathways, it is unlikely that transformation will take place (Burch, 2010; Pelling, 2011). Using Loss and Damage as a lens could broaden perceptions of the causes of, and solutions to, climate change impacts. Framing the issue in the context of local impacts – losses and damages occurring now or in the near future – could make climate change more salient for individuals (Spence, Poortinga, & Pidgeon, 2012). Evidence of the losses and damages being incurred today and forecasting of what could come in the future could change the way in which individuals see themselves and their relationship to the world. As Klein (2014) argues, the world will look different to those who watch the possessions they have worked their entire lives to accumulate float away in the wake of a super storm.

Conclusion

Progress to advance efforts to address loss and damage has been slow and it remains both a technically complicated and politically contentious issue. Loss and Damage is in some sense a newcomer to the UNFCCC agenda, though approaches to address loss and damage were first proposed during the negotiations that led to the UNFCCC in the early 1990s (Roberts & Huq, 2015). More recently, at the 2015 Paris Climate Conference, Loss and Damage became a permanent feature of the global climate change agenda. The Paris Agreement includes areas of cooperation and facilitation with the view to enhancing action and support, including the resilience of communities, livelihoods and ecosystems (UNFCCC, 2016, Article 8). The social turn in climate change research aligns with this opportunity in the UNFCCC architecture to place a developmental reading of climate change more centrally. However, states may fear that, in recognizing the structural causes of vulnerability that lie in local and national decision-making as well as global relations, they could be held more accountable for responses to address vulnerability to climate change impacts. However, as Pope Francis maintained in his encyclical, there is a link between vulnerability to climate change in developing countries and the production and consumption patterns in developed countries. Approaches to avoid, minimize and address loss and damage should be part of a wider movement towards sustainable development. Transformation as liberation acknowledges this possibility.

The IPCC's report on Global Warming of 1.5°C warns that swift and concerted action will be needed to limit global average warming to below 1.5°C (IPCC, 2018). The quicker these actions, the more likely they are to be disruptive to the status quo, which elevates the importance of discussions on what a just transition looks like. The alternative is transformation that is forced on societies when loss and damage occurs. The ambition of enabling social justice through climate change adaptation and wider action to reduce disaster risk is not new, but it has, until recently, been on the fringes of mainstream policy. The UNFCCC Loss and Damage agenda and the synthesis of this burgeoning literature in the IPCC bring transformation centre stage, and with this, opportunities for more emancipatory and liberatory action in the name of climate change adaptation.

As adaptation becomes more integrated into development processes and debates, ongoing and unresolved social struggles will invariably gain prominence. The climate change research community can learn from the processes to better understand the origins of vulnerability to climate change. However, additional research is needed to better understand the social, political and cultural complexities of transformation. As an analogue, where in the past climate change research has called for historical analysis of how society coevolves with climate (see Hulme, 2009), transformation as liberation calls for work that can better understand how individuals and societies cope with deep-rooted social and economic change, and how this can best be steered to move towards, not away from, sustainable development.

Academic research is increasingly engaging with what O'Brien (2012, p. 668) refers to as the 'real adaptive challenge', seen as the questioning of the assumptions, beliefs, values, and interests that have led to the structures, systems and behaviours underlying anthropogenic climate change, and which have also created the conditions for social vulnerability. This calls for a re-framing of climate change, which has until now been mostly characterized as an environmental problem, often neglecting the social, political, cultural and ethical dimensions of the issue (Fook, 2015; O'Brien, 2012). In order to reveal the possibilities for transformation as liberation, more research is needed on the social processes that influence climate change policies and plans, and facilitate transformation (Schipper, Ayers, Reid, Huq, & Rahman, 2014). Scholars and researchers themselves will need to move beyond their comfort zones to grapple with the complexity of adaptation and its role in wider social processes (Fook, 2015; Wise et al., 2014), and may need to transform and change the way in which they see the world. This social turn in responses to avert, minimize and address loss and damage could also usefully explore synergies with the increasing call for evidence and data to monitor national progress towards the SDGs and Sendai Framework for Disaster Risk Reduction, but is also a response to the need for more observational data and sciences, for example epidemiology in climate change impact and adaptation work. Loss and Damage and indeed losses and damages could be an impetus for this deeper, more critical look at how climate change both influences, and is influenced by, broader global, national and sub-national processes.

Disclosure statement

No potential conflict of interest was reported by the authors.

References

ActionAid International. (2011). *What women farmer's need: A blueprint for action*. Johannesburg: Author.

ActionAid International. (2014). Critical stories of change: The fellowship program in Myanmar. Retrieved from http://www.actionaid.org/sites/files/actionaid/critical_story_for_change.pdf

Adger, N. (1999). Social vulnerability to climate change and extremes in coastal Vietnam. *World Development, 27*(2), 249–269.

Adger, N. (2003). Social capital, collective action, and adaptation to climate change. *Economic Geography, 79*(4), 387–404.

Adger, W. N., Dessai, S., Goulden, M., Hulme, M., Lorenzoni, I., Nelson, D. R., … Wreford, A. (2009). Are there social limits to adaptation to climate change? *Climatic Change, 93*, 335–354.

AfDB. (2015). *Empowering African women: An agenda for action, African gender equality index 2015*. Abidjan, Côte d'Ivoire: African Development Bank Group.

Atteridge, A., & Remling, E. (2017). Is adaptation reducing vulnerability or redistributing it? *Wiley Interdisciplinary Reviews: Climate Change, 9*(1), e500. doi:10.1002/wcc.500

Bahadur, A., & Tanner, T. (2012). *Transformation: Theory and practice in climate change and development* (IDS Briefing Note). Brighton: Institute of Development Studies (IDS).

Barnett, J., Evans, L. S., Gross, C., Kiem, A. S., Kingsford, R. T., Palutikof, J. P., … Smithers, S. G. (2015). From barriers to limits to climate change adaptation: Path dependency and the speed of change. *Ecology and Society, 20*(3), 5. doi:10.5751/ES-07698-200305

Béné, C., Newsham, A., Davies, M., Ulrichs, M., & Wood, R. G. (2014). Review article: Resilience, poverty and development. *Journal of International Development, 26*, 598–614. doi:10.1002/jid.2992

Boyd, E., James, R. A., Jones, R. G., Young, H. R., & Otto, F. E. L. (2017). A typology of loss and damage perspectives. *Nature Climate Change, 7*(10), 723. doi:10.1038/nclimate3389

Burch, S. (2010). Transforming enablers of action on climate change: Insights from three municipal case studies in British Columbia, Canada. *Global Environmental Change, 20*, 287–297.

CARE. (2015). *The resilience champions: When women contribute to the resilience of communities in the Sahel through savings and community-based adaptation*. Retrieved from http://www.care-international.org/files/files/Rapport_Resilience_Sahel.pdf

Dodman, D., & Mitlin, D. (2013). Challenges for community-based adaptation: Discovering the potential for transformation. *Journal of International Development, 25*, 640–659.

Dow, K., Berkhout, F., & Preston, B. L. (2013). Limits to adaptation to climate change: A risk approach. *Current Opinion in Environment Sustainability, 5*, 384–391.

Dow, K., Berkhout, F., Preston, B. L., Klein, R. J. T., Midgley, G., & Shaw, M. R. (2013). Commentary: Limits to adaptation. *Nature Climate Change, 3*, 305–307.

Ensor, J., Park, S. E., Hoddy, E. T., & Ratner, B. D. (2015). A rights-based approach on adaptive capacity. *Global Environmental Change, 31*, 38–49. doi:10.1016/j.gloenvcha.2014.12.005

Eriksen, S., Aldunce, P., Bahinipati, C. S., Martins, R. D., Molefe, J. I., Nhemachena, C., … Ulsrud, K. (2011). When not every response to climate change is a good one: Identifying principles of sustainable adaptation. *Climate and Development, 3*, 7–20.

Eriksen, S., Inderberg, T. H., O'Brien, K., & Sygna, L. (2015). Introduction: Development as usual is not enough. In T. H. Inderberg, S. Eriksen, K. O'Brien, & L. Sygna (Eds.), *Climate change adaptation and development: Transforming paradigms and practices* (pp. 1–18). London: Routledge.

Fenton, A., Tallontire, A., & Paavola, J. (2016). *Autonomous adaptation to riverine flooding in Satkira District, Bangladesh: Insights for transformation* (Sustainability Research Institute Paper No. 99/Centre for Climate Change Economics and Policy Working Paper No. 283). Retrieved from https://www.see.leeds.ac.uk/fileadmin/Documents/research/sri/workingpapers/SRIPs-99.pdf

Feola, G. (2015). Societal transformation in response to global environmental change: A review of emerging concepts. *Ambio, 44,* 376–390.

Folke, C. (2006). Resilience: The emergence of a perspective for social-ecological systems analyses. *Global Environmental Change-Human and Policy Dimensions, 16*(3), 253–267.

Folke, C., Carpenter, S. R., Walker, B., Scheffer, M., Chapin, T., & Rockström, J. (2010). Resilience thinking: Integrating resilience, adaptability and transformability. *Ecology and Society, 15*(4), 20. Retrieved from http://www.ecologyandsociety.org/vol15/iss4/art20/

Fook, T. C. T. (2015). Transformational processes for community-focused adaptation and social change: A synthesis. *Climate and Development, 9*(1), 5–21. doi:10.1080/17565529.2015.1086294

Francis. (2015). Encyclical letter Laudato Si' of the Holy Father Francis on care for our common home. Retrieved from http://w2.vatican.va/content/francesco/en/encyclicals/documents/papa-francesco_20150524_enciclica-laudato-si.html

Freire, P. (1970). *Pedagogy of the oppressed.* London: Penguin.

Godfrey-Wood, R., & Naess, L. O. (2016). Adapting to climate change: Transforming development. *IDS Bulletin, 47*(2), 49–62.

Gunderson, L. H., & Holling, C. S. E. (2002). *Panarchy: Understanding transformations in human and natural systems.* Washington, DC: Island Press.

Gunderson, L. H., Holling, C. S., & Light, S. S. (1995). *Barriers and bridges to the renewal of ecosystems and institutions.* New York, NY: Columbia University Press.

Hashemi, S. M., & Umaira, W. (2011). *New pathways for the poorest: graduation from the model from BRAC* (CSP Research Report 10). Sussex: Centre for Social Protection, Institute of Development Studies.

Holling, C. S. (1986). The resilience of terrestrial ecosystems: Local surprise and global change. In W. C. Clark & R. E. Munn (Eds.), *Sustainable development of the Biosphere* (pp. 292–317). London: Cambridge University Press.

Hulme, M. (2009). *Why we disagree about climate change.* Cambridge: Cambridge University Press.

IPCC. (2012). Managing the risks of extreme events and disasters to advance climate change adaptation. In C. B. Field, V. Barros, T. F. Stocker, D. Qin, D. J. Dokken, K. L. Ebi, … P. M. Midgley (Eds.), *A special report of working groups I and II of the Intergovernmental Panel on Climate Change.* Cambridge: Cambridge University Press.

IPCC. (2014). Climate change 2014: Synthesis report. In CoreWriting Team, R. K. Pachauri, & L. A. Meyer (Eds.), *Contribution of working groups I, II and III to the fifth assessment report of the Intergovernmental Panel on Climate Change* (151 pp.). Geneva: IPCC.

IPCC. (2018). Summary for policymakers. In V. Masson-Delmotte, P. Zhai, H. O. Pörtner, D. Roberts, J. Skea, P. R. Shukla, … T. Waterfield (Eds.), *Global warming of 1.5°C. An IPCC special report on the impacts of global warming of 1.5°C above pre-industrial levels and related global greenhouse gas emission pathways, in the context of strengthening the global response to the threat of climate change, sustainable development, and efforts to eradicate poverty* (32 pp.). Geneva: World Meteorological Organization.

Jakku, E., Thornburn, P. J., Marshall, N. A., Dowd, A. M., Howden, S. M., Mendham, E., … Brandon, C. (2016). Learning the hard way: A case study of an attempt at agricultural transformation in response to climate change. *Climatic Change, 137,* 557–574.

Joshi, D., Platteeuw, J., Singh, J., & Teoh, J. (2018). Watered down? Civil society organizations and hydropower development in the Darjeeling and Sikkim regions, Eastern Himalaya: A comparative study. *Climate Policy, 19*(1), 63–77.

Kates, R. W., Travis, W. R., & Wilbanks, T. J. (2012). Transformational adaptation when incremental adaptations to climate change are insufficient. *Proceedings of the National Academy of Science, 109*(19), 7156–7161.

Klein, N. (2014). *This changes everything: Capitalism vs. The climate.* Toronto: Alfred A. Knopf Canada.

Klein, R. J. T., Midgley, G. F., Preston, B. L., Alam, M., Berkhout, F. G. H., Dow, K., & Shaw, M. R. (2014). Adaptation opportunities, constraints, and limits. In C. B. Field, V. R. Barros, D. J. Dokken, K. J. Mach, M. D. Mastrandrea, T. E. Bilir, … L. L. White (Eds.), *Climate change 2014: Impacts, adaptation, and vulnerability. Part A: Global and sectoral aspects. Contribution of working group II to the fifth assessment report of the Intergovernmental Panel on Climate Change* (pp. 899–943). Cambridge: Cambridge University Press.

Kreft, S., Warner, K., Harmeling, S., & Roberts, E. (2013). Framing the loss and damage debate: A conversation starter by the loss and damage in vulnerable countries initiative. In O. C. Ruppel, C. Roschmann, & K. Ruppel-Schlichting (Eds.), *Climate change: International law and global governance* (Vol. II: Policy, Diplomacy and Governance in a Changing Environment). Munich: Novos.

Manuel-Navarrete, D., & Pelling, M. (2015). Subjectivity and the politics of transformation in response to development and environmental change. *Global Environmental Change, 35,* 558–569.

Mapfumo, P., Onyango, M., Honkponou, S. K., El Mzouri, E. H., Githeko, A., Rabeharisoa, L., … Agrawal, A. (2015). Pathways to transformational change in the face of climate impacts: An analytical framework. *Climate and Development, 9*(5), 439–451. doi:10.1080/17565529.2015.1040365

Marshall, N. A., Park, S. E., Adger, W. N., Brown, K., & Howden, S. M. (2012). Transformational capacity and the influence of place and identity. *Environmental Research Letters, 7,* 034022. doi:10.1088/1748-9326/7/3/034022

Matyas, D., & Pelling, M. (2014). Positioning resilience for 2015: The role of resistance, incremental adjustment and transformation in disaster risk management policy. *Disasters, 30*(S1), S1–S18.

Mitlin, D. (2012). Lessons from the urban poor: Collective action and the rethinking of development. In M. Pelling, D. Manuel-Navarrete & M. Redclift (Eds.), *Climate change and the crisis of capitalism: A change to reclaim self, society and nature* (pp. 85–98). London: Routledge.

Moore, M.-L., Tjornbo, O., Enfors, E., Knapp, C., Hodbod, J., Baggio, J. A., … Biggs, D. (2014). Studying the complexity of change: Toward an analytical framework for understanding deliberate social-ecological transformations. *Ecology and Society, 19*(4), 45.

Nalau, J., & Handmer, J. (2015). When is transformation a viable policy alternative. *Environmental Science and Policy, 54,* 349–356.

Nelson, D. R., Adger, W. N., & Brown, K. (2007). Adaptation to environmental change: Contributions of a resilience framework. *Annual Review of Environment and Resources, 32,* 395–419. doi:10.1146/annurev.energy.32.051807.090348

O'Brien, K. (2012). Global environmental change II: From adaptation to deliberate transformation. *Progress in Human Geography, 36,* 667–676.

O'Brien, K. (2017). Climate change adaptation and social transformation. *The International Encyclopedia of Geography.* doi:10.1002/9781118786352.wbieg0987

O'Brien, K., Pelling, M., Patwardhan, A., Hallegatte, S., Maskrey, A., Oki, T., … Yanda, P. Z. (2012). Toward a sustainable and resilient future. In C. B. Field, V. Barros, T. F. Stocker, D. Qin, D. J. Dokken, K. L. Ebi, … P. M. Midgley (Eds.), *Managing the risks of extreme events and disasters to advance climate change adaptation. A special report of working groups I and II of the Intergovernmental Panel on Climate Change (IPCC)* (pp. 437–486). Cambridge: Cambridge University Press.

Olsson, P., Galaz, V., & Boonstra, W. J. (2014). Sustainability transformations: A resilience perspective. *Ecology and Society, 19*(4), 1.

O'Neil, S. J., & Handmer, J. (2012). Responding to bushfire risk: The need for transformative adaptation. *Environmental Research Letters, 7,* 1–7. doi:10.1088/1748-9326/7/1/014018

Opondo, D. O. (2013). Erosive coping after the 2011 floods in Kenya. *International Journal of Global Warming, 5*(4), 452–466.

Paprocki, K. (2015, September). Anti-politics of climate change: Depoliticisation of climate change undermines historic reasons that made Bangladesh vulnerable to it. *Himalayan Southasian.* Retrieved from http://www.kasiapaprocki.com/uploads/5/5/8/1/55818931/paprocki_anti-politics_of_climate_change_himal.pdf.

Park, S. E., Marshall, N. A., Jakku, E., Dowd, A. M., Howden, S. M., Mendham, E., & Fleming, A. (2012). Informing adaptation responses to climate change through theories of transformation. *Global Environmental Change, 22,* 115–126. doi:10.1016/j.gloenvcha.2011.10.003

Parry, M. (2009). Climate change is a development issue, and only sustainable development can confront the challenge. *Climate and Development, 1*(2009), 5–9.

Parsons, M., & Nalau, J. (2016). Historical analogies as tools in understanding transformation. *Global Environmental Change, 38,* 82–96.

Pearson, L. J., & Pearson, C. J. (2012). Societal collapse or transformation, and resilience. *PNAS, 109*(30), E2030–E2031. doi:10.1073/pnas.1207552109

Pelling, M. (2011). *Adaptation to climate change: From resilience to transformation.* New York, NY: Routledge.

Pelling, M. (2014). *Pathways to transformation: Disaster risk management to enhance development goals.* Background paper prepared for the Global Assessment Report on Disaster Risk Reduction. Retrieved from http://www.preventionweb.net/english/hyogo/gar/2015/en/bgdocs/Pelling,%202014.pdf

Pelling, M., & Dill, K. (2010). Disaster politics: Tipping points for change in the adaptation of sociopolitical regimes. *Progress in Human Geography, 34*(1), 21–37.

Pelling, M., & Manuel- Navarrete, D. (2011). From resilience to transformation: The adaptive cycle in two Mexican urban centers. *Ecology and Society, 16*(2), 11.

Pelling, M., O'Brien, K., & Matyas, D. (2015). Adaptation and transformation. *Climatic Change, 128*(1–2), 1–17. doi:10.1007/s10584-014-1303-0

Preston, B. L., Dow, K., & Berkhout, F. (2013). The climate adaptation frontier. *Sustainability, 5,* 1011–1035.

Räthzel, N., & Uzzell, D. (2011). Trade unions and climate change: The jobs versus environment dilemma. *Global Environmental Change, 21,* 1215–1223. doi:10.1016/j.gloenvcha.2011.07.010

Ribot, J. (2010). Vulnerability does not fall from the Sky: Toward a multiscale, pro-poor climate policy. In R. Mearns & A. Norton (Eds.), *Social dimensions for climate change: Equity and vulnerability in a warming world* (pp. 47–74). Washington, DC: The World Bank.

Rickards, L. (2013). Transformation is adaptation. *Nature Climate Change, 3,* 690.

Rickards, L., & Howden, S. M. (2012). Transformational adaptation: Agriculture and climate change. *Crop and Pasture Science, 63,* 240–250.

Rippke, U., Ramirez-Villegas, J., Jarvis, A., Vermeulen, S. J., Parker, L., Mer, F., … Howdern, M. (2016). Timescales of transformational climate change adaptation in sub-Saharan African agriculture. *Nature Climate Change, 6,* 605–610.

Roberts, E., & Huq, S. (2015). Coming full circle: The history of loss and damage under the UNFCCC. *International Journal of Global Warming, 8*(2), 141.

Roberts, E., & Pelling, M. (2018). Climate change-related loss and damage: Translating the global policy agenda for national policy processes. *Climate and Development, 10*(1), 4–17.

Roberts, E., van der Geest, K., Warner, K., & Andrei, S. (2014). Loss and damage: When adaptation is not enough. *Environmental Development, 11,* 219–227.

Rosenzweig, C., & Solecki, W. (2014). Hurricane Sandy and adaptation pathways in New York: Lessons from a first-responder city. *Global Environmental Change, 28,* 395–408. doi:10.1016/j.gloenvcha.2014.05.003

Schipper, E. L. F., Ayers, J., Reid, H., Huq, S., & Rahman, A. (2014). *Community based adaptation to climate change: Scaling it up.* London: Routledge.

Schlitz, M. M., Vieten, C., & Miller, E. M. (2010). Worldview transformation and the development of social consciousness. *Journal of Consciousness Studies, 17*(7–8), 18–36.

Smith, A., Stirling, A., & Berkhout, F. (2005). The governance of sustainable socio-technical transitions. *Research Policy, 34*(10), 1491–1510. doi:10.1016/j.respol.2005.07.005

Spence, A., Poortinga, W., & Pidgeon, N. (2012). The psychological distance of climate change. *American Psychologist, 66*(4), 265–276.

Stafford, M. S., Horrocks, L., Harvey, A., & Hamilton, C. (2011). Rethinking adaptation for a 4C world. *Philosophical Transactions of the Royal Society A, 369,* 196–216.

Surminski, S., & Lopez, A. (2015). Concept of loss and damage of climate change – a new challenge for climate decision-making? A climate science perspective. *Climate and Development, 7*(3), 267–277.

Thornton, T. F., & Camberti, C. (2013). Synergies and trade-offs between adaptation, mitigation and development. *Climatic Change*, *146*(1), 5–18. doi:10.1007/s10584-013-0884-3

Tschakert, P., & Dietrich, K. A. (2010). Anticipatory learning for climate change adaptation and resilience. *Ecology and Society*, *15*(2), 11 Retrieved from http://www.ecologyandsociety.org/vol15/iss2/art11/

UNFCCC. (2012). A literature review on the topics in the context of thematic area 2 of the work programme on loss and damage: A range of approaches to address loss and damage associated with the adverse effects of climate change. FCCC/SBI/2012/INF.14.

UNFCCC. (2014). Report of the conference of the parties on its nineteenth session, held in Warsaw from 11 to 23 November 2013 FCCC/CP/2013/10/Add.1.

UNFCCC. (2016). Report of the conference of the parties on its twenty-first session, held in Paris from 30 November to 13 December 2015. FCCC/CP/2015/10/Add.1.

UNFCCC. (2017). The five-year rolling workplan of the Executive Committee of the Warsaw international mechanism for loss and damage associated with climate change impact. Retrieved from https://unfccc.int/files/adaptation/groups_committees/loss_and_damage_executive_committee/application/pdf/draft-five-year-rolling-workplan-12-oct.pdf

Walker, B., Holling, C. S., Carpenter, S. R., & Kinzig, A. (2004). Resilience, adaptability and transformability in social-ecological systems. *Ecology and Society*, *9*, 5.

Warner, K., & van der Geest, K. (2013). Loss and damage from climate change: Local level evidence from nine vulnerable countries. *International Journal of Global Warming*, *5*(4), 367–386.

Warner, K., van der Geest, K., Kreft, S., Huq, S., Harmeling, S., Kusters, K., & de Sherbinin, A. (2012). *Evidence from the frontlines of climate change: Loss and damage to communities despite coping and adaptation. Loss and damage in vulnerable countries initiative* (Policy Report No. 9)..

Westley, F., Olsson, P., Folke, C., Homer-Dixon, T., Vredenburg, H., Loorbach, D., ... van der Leeuw, S. (2011). Tipping toward sustainability: Emerging pathways of transformation. *Ambio*, *40*(7), 762–780.

Westley, F. R., Tiorho, O., Schultz, L., Olsson, P., Folke, C., Crona, B., & Bodin, O. (2013). A theory of transformative agency in linked social ecological systems. *Ecology and Society*, *18*(3), 27.

Wilson, S., Pearson, L. J., Kashima, Y., Lusher, D., & Pearson, C. (2013). Separating adaptive maintenance (resilience) and transformative capacity of social-ecological systems. *Ecology and Society*, *18*(1), 22.

Wise, R. M., Fazey, I., Stafford Smith, M., Park, S. E., Eakin, H. C., Archer Van Garderen, E. R. M., & Campbell, B. (2014). Reconceptualising adaptation to climate change as part of pathways of change and response. *Global Environmental Change*, *28*, 325–336. doi:10.1016. j.gloenvcha.2013.12.002

Wisner, B., Cannon, T., Davis, I., & Blaikie, P. (2004). *At risk*. London: Routledge.

Ziervogel, G., Pelling, M. A., Cartwright, A., Chu, E., Deshpande, T., Harris, L., ... Zweig, P. (2017). Inserting rights and justice into urban resilience: A focus on everyday risk. *Environment and Urbanization*, *29*(1), 123–138.

Index

Note: **Bold** page numbers refer to tables; *italic* page numbers refer to figures and page numbers followed by "n" denote endnotes.